SHIP TO SHORE, INC.

SHIP TO SHORE I

CARIBBEAN CHARTER YACHT RECIPES

Publisher and Editor .... Capt. Jan Robinson

Associate Editor .... Cheryl Anne Fowler

Associate Publisher .... Waverly H. Robinson

Title by .... F. Marion Meginnis

Artist................................. Celia A. Flock

SHIP TO SHORE, INC.

10500 Mt. Holly Road
Charlotte, NC 28214-9219
704-394-2433 Phone 704-392-4777 Fax
E-mails: CapJan@aol.com

P.O. Box 10898
St. Thomas, US Virgin Islands, 00801
340-775-6295 Phone/Fax

To order additional copies or a *free* catalog
1-800-338-6072

SHIP TO SHORE I
COOKBOOK

Copyright 1983 by Ship to Shore, Inc

First printing: October, 1983
Second printing: February, 1984
Third printing: December, 1984
Fourth printing: March, 1987
Fifth printing: October, 1987
Sixth printing: October, 1990
Seventh printing: November, 1991
Eighth printing: September, 1993
Ninth printing: August, 1995
Tenth printing: August, 1997
Eleventh printing: January, 1999

Printed in the United States of America

ISBN 0-9612686-0-3

FOREWORD

Discover the secrets ...
The spray of saltwater ...
The romance of island hopping ...
The heady breezes of the Caribbean. All this can fuel a hearty
appetite.

Great changes, that have ▓▓▓▓ tionized the role of the charter yacht
chef, have taken place in ▓▓▓▓ ter yacht community. In the past, the
majority of charter guests ▓▓▓▓ were simple - to enjoy the beauty of the
islands, relax in the comf▓ ▓ yacht and be given enough nourish-
ment to sustain them from ▓▓▓▓ port.

As the demand for more ▓▓▓▓ grew, so did the demand for more
sophisticated attention to th▓ ▓▓▓ ry needs and requirements of charter
guests. Guests are no longe ▓▓▓▓ t with a menu consisting of peanut
butter and jelly sandwiches ▓▓▓▓ l by the first mate. Real progress has
been made since the days of iron ships, wooden men and salt pork,
chased with old hardtack.

Today, charter yacht chefs are well trained professionals. They possess
the ability to create exciting meals for the most discriminating guests.
On a daily basis, they create dining experiences equal to that of the
world's finest restaurants. Each dish reflects a zest for providing dining
elegance with a minimum of effort. From elegant appetizers to
mouthwatering main dishes and luscious desserts, you'll find meal
planning a snap!

Ship to Shore I is a fascinating, and intriguing collection of closely
guarded gourmet recipes from luxury yacht galleys. Jan Robinson, as
well as other Caribbean charter yacht chefs, now share their treasured
recipes for healthful, elegant and easy meals you can make at home.
With a galley smaller than most home kitchens, and with no conve-
niences nearby, the chefs create fabulous meals morning, noon, and
night.

In addition to the innovative recipes utilizing readily available ingredients, this convenient and easy-to-read collection includes preparation
and cooking times for easy meal planning, valuable hints, suggested
appetizers, accompaniments and garnishes. Also personal anecdotes
and fun to read stories about the chefs and their yachts.

So weigh anchor and set sail on this culinary cruise. Bon appetit and bon
voyage!

TECHNIQUES TO HELP YOU GET INTO HEALTHY COOKING

Adjusting to a lifestyle of healthier eating is really a simple matter of using some smart easy cooking techniques. Below is an introduction to preparing healthy meals with ingredient substitutions to reduce sodium, fat, cholesterol, sugar, and calories, plus adding fiber to one's diet. You can apply these techniques to the *Ship to Shore* recipes and many of your own favorites.

To Reduce Fat and Cholesterol:
- Use nonfat, and no or low cholesterol dairy products.
- Replace one cup whole milk with one cup nonfat milk.
- Use nonfat mayonnaise or nonfat yogurt in place of mayonnaise.
- Use nonfat yogurt with fresh chives in place of sour cream on baked potatoes.
- Use fat-free cottage cheese in place of mayonnaise or sour cream in dips.
- Season foods with salsas rather than butter-based sauces.
- Use skimmed evaporated milk instead of cream.
- Use crushed high-fiber cereal flakes in place of buttered bread crumbs.
- Remove skin from chicken and fish.
- Substitute two egg whites, or 1/4 cup egg substitute for each whole egg.
- Broil, bake, roast, boil, or steam in order not to add any fat. Saute' in a nonstick skillet or wok using a small amount of water or broth instead of butter or oil.
- Make gravy from bouillon, stock, or wine and herbs.
- Refrigerate stocks, soups, or stews so the fat can float and condense on the top; skim off before reheating and serving.
- Try the turkey-based version for sausage, bacon, or ham.
- Select lean cuts of meat. Look for words like "loin" or "round" in the name.
- Substitute ground turkey or chicken breast for ground beef or pork.

To Reduce Sugar:
- Choose canned fruits in their own juice or water-packed.
- Scan product labels for words on ingredient list such and dextrose, sucrose, corn syrup solids, fructose or high fructose corn sweetener. These are all words for sugar.

Saving The Vitamins in Cooked Vegetables:
- Leave the peels on.
- Cut, slice, dice, etc. after cooking.
- Steam, stir-fry, microwave, or boil in a little water; cover while cooking.
- Cook until tender but crunchy.

To Reduce Sodium:
- Use fresh or frozen ingredients in recipes. If you must use canned, drain and rinse in fresh water before using.
- Use low-sodium cheese.
- Use the following flavor enhancers:
  Fresh herbs and spices
  Dried herbs and spices
  No-salt or low-sodium vegetable or tomato juice to flavor soups and stews.
  Grated lemon, lime, or orange peel to flavor chicken, fish, salads, and vegetables, or one or two tablespoons of apple or orange juice concentrate.
- Make homemade stocks with leftover vegetables rather than broth bases or bouillon,
- Spray popcorn lightly with a nonstick cooking spray to makeseasoning adhere. Flavor with chili, garlic or onion powder.

To Add Fiber:
- Substitute whole-wheat flour (half the quantity) for all-purpose flour when baking.
- Add 2 tablespoons wheat germ or unprocessed bran for each cup of flour.
- Use brown rice, whole wheat, or vegetable-base noodles and pastas.
- Serve fresh fruits and vegetables with their skin on.
- Add sunflower, sesame, poppy seeds for crunch and taste.
- Use wheat germ, unprocessed bran, or uncooked oats in salads, casseroles, and muffins or as a light coating for baked chicken or fish.
- Add dried fruits to recipes.
- Make your own high-fiber granola with oats, seeds, nuts, and dried fruits.
  (Nuts are rich in fat, so do not go overboard.)

TABLE OF CONTENTS

BREAKFAST AND BREADS

TABLE OF EQUIVALENTS
UNITED STATES AND METRIC

U.S.	EQUIVALENTS	METRIC volume-milliliters
Dash	Less than 1/8 tsp.	
1 teaspoon	60 drops	5 ml
1 Tablespoon	3 teaspoons	15 ml.
2 Tablespoons	1 fluid ounce	30 ml.
4 Tablespoons	1/4 cup	60 ml.
5-1/3 Tablespoons	1/3 cup	80 ml.
6 Tablespoons	3/8 cup	90 ml.
8 Tablespoons	1/2 cup	120 ml.
10-2/3 Tablespoons	2/3 cup	160 ml.
12 Tablespoons	3/4 cup	180 ml.
16 Tablespoons	1 cup or 8 oz.	240 ml.
1 cup	1/2 pint or 8 oz.	240 ml.
2 cups	1 pint	480 ml.
1 pint	16 oz.	480 ml.
1 quart	2 pints	960 ml.
2.1 pints	1.05 quarts	1 liter
2 quarts	1/2 gallon	1.9 liter
4 quarts	1 gallon	3.8 liters

		weight-grams
1 ounce	16 drams	28 grams
1 pound	16 ounces	454 grams
1 pound	2 cups liquid	
1 kilogram	2.20 pounds	

AVOCADO EGGS

preparation time: 15-20 minutes   'Ann Glenn
cooking time : 10 minutes (Microwave Oven)   *Encore*
serves: 6-8

6-8 hard boiled eggs
3-4 English muffins, split, buttered
   and broiled
2 Tblsp. butter
1 large onion, chopped

1 chile, or 1 can chilies, seeded &
   minced
2 Tblsp. flour
1 cup light cream (Pet milk is fine)
2 ripe avocados, seeded & mashed

Combine butter, onion and chilies in glass bowl. Microwave until limp, 2-3 minutes. Add flour and milk (season with salt and pepper). Cook 1 minute and stir to make cream sauce. Add mashed avocados and heat again, 1-2 minutes. Peel and slice hard-boiled eggs. Add 3 eggs to avocado sauce and heat through, about 1 minute. Add another 3 eggs and heat through, then last 4 and heat. *Can be served buffet style, with separate plates for muffins, sliced ham and avocado sauce, or it can be "stacked" to build open-faced sandwiches. In either case, children, young parents and grandparents can be relied upon to recognize "GREEN EGGS AND HAM"! Also serve a Blintz Bubble Ring and ham slices with syrup.*

SHERRIED HAM AND EGGS

preparation time: 20 minutes   *Ann Glenn*
cooking time: 20-30 minutes   *Encore*
serves 8

6-8 oz. cooked, sliced ham
1 box herb seasoned croutons
8-12 eggs
8 oz. light cream, half 'n' half or
   Pet milk

4 Tblsp. dry sherry
¼ tsp. cayenne pepper
1 tsp. Worcestershire sauce
2 cups Swiss or Jarlsberg cheese,
   grated

Grease a 9x13" serving dish well. Cut the sliced ham into bite-sized pieces and cover the bottom and sides of the pan over-lapping slightly. Layer the package of croutons and spread evenly. Carefully crack the eggs over the croutons, hoping the yolk will not break and dribble away among the croutons. In a small pitcher, combine sherry, cream, Worcestershire sauce and cayenne. Drizzle a tablespoon of this sauce over each egg and place remaining sauce on the stove to warm before taking to the table. Spread grated cheese over the top. Bake at 375° for 20 minutes before testing to see if egg yolks are cooked firm enough. *Serve with Island Coffee Cake.*

CHRISTIAN'S SPECIAL

preparation time: 5 minutes   Linda Richards
cooking time: 20 minutes   Tao Mu
serves 8

4-5 slices bacon   ½ cup white wine
½ small onion, finely chopped   16 eggs
2 cans cream of mushroom soup   16 slices buttered toast
1 (14 oz.) can evaporated milk

Chop bacon, fry until brown and drain except for 1 Tblsp. grease. Add
chopped onion and saute over low heat until transparent. Increase heat to
medium, add soup, milk and wine to bacon and onion and bring just to a
boil (do not boil). Poach the eggs in the sauce, occasionally spooning the
sauce over the eggs to cook them, being careful not to break them. *Serve
on buttered toast — 2 per person. May be garnished with parsley, thin
slice tomato, etc.*

UN-BORING BREAKFAST

preparation time: 15 minutes   Different Drummer
cooking time: 10 minutes
serves 4

1 small green pepper, chopped   Pepper
1 small onion, chopped   Parsley
1 cup ham, chopped   Paprika
8 lightly beaten eggs   English muffins, toasted and
Garlic powder   buttered
Salt

Saute green pepper, onion, and ham only until the vegetables are warm,
but not limp. Season the eggs with garlic powder, salt, pepper and parsley. Add the seasoned eggs to the vegetables. Stir gently until creamy.
*Serve on large platter with paprika and parsley sprinkles surrounded by
toasted, buttered English muffins.*

• *When you burn yourself in the galley/kitchen, vanilla will help ease
the pain (apply it; don't drink it)*

BRUNCH CASSEROLE

preparation time: 15 minutes
cooking time: 10-15 minutes
serves 6

Pamela McMichael
Pieces of Eight

Salt and pepper
6 hard-boiled eggs, sliced
1 lb. hot bulk sausage

1½ cups sour cream
½ cup dry bread crumbs
1½ cups cheddar cheese, grated

Place eggs in buttered casserole and season to taste. Cook sausage, drain and sprinkle over eggs. Pour sour cream over sausage. Combine crumbs and cheese. Sprinkle over casserole. Place in oven to heat thoroughly and brown top under broiler. *A favorite with everyone who tries it.*

EGGS JEFFERY

preparation time: 15 minutes
cooking time: 5 minutes
serves 6

Kandy Popkes
Wind Song Oregon

1 small chopped onion
½ green pepper, chopped
1½ cups sharp cheddar cheese,
    grated
6 eggs

Sauce:
½ cup ketchup
⅛ cup BBQ sauce
¼ tsp. Worcestershire
6 drops Tabasco

Saute onions and green pepper in a little oil. In frying pan put 6 eggs on low heat. Sprinkle with sauteed onion, green pepper, sauce and cheese. Cover and cook till cheese is melted. *Serve and garnish with orange slices and parsley.*

ITALIAN OMELETTE

preparation time: 5 minutes
cooking time: 10 minutes
serves 8

Adriane Biggs
Taza Grande

3 Tblsp. butter
2 lbs sweet Italian sausage, sliced
1 large green pepper, chopped
1 medium onion, chopped
1 clove garlic, crushed

1 medium tomato, skinned and
    chopped
¼ tsp. crushed red pepper
1 tsp. salt
¼ tsp. black pepper
8 eggs, beaten

In covered skillet, melt butter. Saute sausage, green pepper, onion and garlic until tender. Add tomato and crushed red pepper and saute 1 minute longer. Add eggs, salt and pepper. Cover and cook over low heat until done.

SPANISH OMELETTE

preparation time: 30 minutes   Samantha Lehman
cooking time: 30 minutes   Lady Sam
serves 6

1 large can stewed tomatoes	Garlic powder
1 small can tomato paste	1 small can diced chiles
Chopped onion	Cornstarch

Above is for the sauce to cover your favorite omelette; preferably denver (green pepper, ham, onion, and cheese), cheese or cheese and onion. I have been most successful with all eggs (except fried) by adding Parmesan cheese and always garlic.

In a saucepan pour stewed tomatoes, tomato paste, onion, garlic powder (lightly) and diced chiles (add seeds according to your pallet). Heat all of the above, then thicken sauce with cornstarch (or Chinese rice flour). Mix and stir until right consistency. Set at low heat, make your omelettes.

CHICKEN LIVER OMELETTE

preparation time: 15 minutes   Launa Cable
cooking time: 20 minutes   Antiquity
serves 4-6

4 Tblsp. butter	1 tsp. salt
1 lb. chicken livers, cut in half	¼ tsp. pepper
1 tsp. salt	2 Tblsp. butter
¼ tsp. white pepper	2 Tblsp. snipped chives
½ tsp. oregano	¼ cup sour cream
6 eggs	1 tomato, peeled, seeded and
¼ cup cream cheese, melted	chopped

In large skillet, melt 4 Tblsp. butter. Add livers, 1 tsp. salt, white pepper, and oregano. Saute livers about 5 minutes. Cover and keep warm. Beat eggs, 1 tsp. salt, and pepper. Melt 2 Tblsp. butter, add egg mixture. Slide spatula around edge of pan carefully lifting egg so uncooked eggs will flow underneath. Cook until set. Carefully place on warm platter. Arrange livers and chopped tomato over top of omelette. Spoon mixture of sour cream and cream cheese on top. Sprinkle with chives. Garnish sides with parsley.

NO FUSS LANGOSTINO SOUFFLE

preparation time: 5 minutes   Samantha Lehman
cooking time: 45 minutes   Lady Sam
serves 6

2 eggs (per person)   Black pepper, to taste
Chopped onion   1 tsp. dried red pepper
Dash parsley (or flakes) for color   Parmesan cheese (¼ cup min.)
Garlic powder, to taste   1 pkg. Langostino (about 1 cup)

Mix all ingredients (except langostino) in blender. Pour into greased baking dish. Sprinkle langostino chunks on top. Place in oven at 400° approximately 45 minutes until done (use a fork lightly in center until very little juice). *Serve with chilled orange juice, champagne and toasted English muffins.*

SCRAMBLED EGGS ON BAKED POTATO BOATS

preparation time: 35 minutes   Jane Marriott
cooking time: 20 minutes   Jan Pamela II
serves 8

8 Tblsp. butter   8 large Idaho potatoes, baked and
1½ Tblsp. flour   cooled (can be done the night
⅔ cup sour cream   before)
16 eggs   ½ cup melted butter
Salt and pepper to taste

Cut each potato in half lengthwise and scoop out pulp leaving a ¼ inch shell. Put shells on a baking sheet and brush the insides with melted butter. Bake for 20 minutes at 425° while you make the eggs.

In a small pan, melt 1½ Tblsp. butter, stir in the flour to form a "roux". Off the heat, blend in the sour cream. Cook for a few minutes, stirring until thick and smooth, then set aside.

Beat together the eggs, salt and pepper to taste, then scramble with the remaining 1½ Tblsp. butter in the usual way being careful not to overcook the eggs. When they are set but still soft, stir in the sour cream mixture. This method of scrambling eggs enables them to be held in a warm place for a long time without either separating or going hard.

Fill the potato boats with the eggs and serve with crisp-fried bacon and sausage and/or all of the following side dishes:

grated yellow cheese; crumbled blue cheese; chopped green onions; chopped black olives; cooked baby shrimp; red or black caviar

NO PEEK-E-EGGS

preparation time: 10 minutes   Samantha Lehman
cooking time: 10-15 minutes   Lady Sam
serves 6

12 eggs (2 eggs per person)
1 green pepper (small or medium),
   chopped
1 onion (small) chopped
1 tomato (average) chopped
¼ cup sharp cheese (more if you
   like cheesy flavor)
Margarine (butter burns in
   cooking)

Garlic powder, to taste (don't be
   too stingy)
Parsley flakes, for color
1 tsp. dried red pepper
Diced ham, tuna, shrimp, bacon,
   pork or whatever (use your
   leftovers if any)

Take all the above (except margarine and meat or fish) and place in your blender or bowl and whip. In your deep skillet, heat your margarine (remember butter burns). Pour ingredients into heated skillet, top with your meat or fish, cover and let cook at medium flame until ready (use your own judgment as to dry or just moist enough), about 10 minutes or less. Will be full and fluffy. *Cut into pie sections and serve.*

FRENCH CREAMED EGGS WITH LEEKS

prepration time: 10 minutes   Kathleen McNally
cooking time: 20-25 minutes   Orka
serves 8-10

3 Tblsp. butter
4 medium leeks (white part only),
   thinly sliced (4 to 4½ cup)
2 Tblsp. butter (¼ stick)
20 eggs, beaten to blend
8 oz. cream cheese (cut into small
   pieces at room temperature)

1 Tblsp. fresh mint or 1 tsp. dried
Salt
Freshly ground pepper
Fresh mint sprigs (garnish)

Sweat leeks in 3 Tblsp. butter over low heat until soft and transparent, about 20 minutes, uncover and continue cooking if necessary until all liquid evaporates. Remove leeks from skillet. Melt 2 Tblsp. butter in same skillet over low heat. Add eggs, cream cheese, mint, salt and pepper and cook stirring with whisk or fork until eggs begin to set. Stir in leeks and cook until mixture forms soft curds. Taste and adjust seasoning. Turn out onto heated platter. Garnish with mint.

• *Hold eggs together while poaching by adding a few drops of vinegar or lemon juice to water.*

RANCH-STYLE EGGS

preparation time: 25 minutes   *Shirley Fortune*
cooking time: 5 minutes   *Maranatha*
serves 4

Mexican Sauce (below)	**8 eggs**
Vegetable Oil	**Salt and pepper**
8 four-inch tortillas-flat kind	**1 Cup shredded Monterey Jack**
Vegetable Oil	** cheese (4 ounces)**

Prepare Mexican Sauce. Heat oil (⅛ inch) in 6 or 8-inch skillet until hot. Cook tortillas until crisp and light brown, about 1 minute on each side. Drain on paper towels, keep warm.

Heat oil (⅛ inch) in 12-inch skillet until hot. Break each egg into measuring cup or saucer; carefully slip 4 eggs, 1 at a time into skillet. Immediately reduce heat. Cook slowly spooning oil onto eggs until whites are set and a film forms over the yolks. Or, turn eggs over gently when whites are set and cook to desired doneness. Sprinkle with salt and pepper. Repeat with remaining eggs.

Spoon 1 tablespoon sauce over each tortilla; place 1 egg on each. Spoon sauce over white of egg; sprinkle yolk with cheese.

MEXICAN SAUCE

1 medium onion chopped	**1 tablespoon vegetable oil**
½ green pepper, chopped	**2 cups chopped ripe tomatoes**
1 clove garlic, finely chopped	**¼ to ½ cup chopped green chilies**

EGGS TAURMINA

preparation time: 10 minutes   *Capt. Joe Garrison*
cooking time: 8-10 minutes   *Bluejacket*
serves 8-12

2 cans cream of mushroom soup	**Salt**
1 soup can of milk	**Pepper**
12 eggs	**Celery salt**
1 medium onion, finely chopped	**Oregano**
1 green pepper, finely chopped	

Saute onion and pepper in butter until soft. Mix all ingredients in large (8x14) casserole dish until well blended. Add salt, pepper, celery salt and dash of oregano. Evenly space eggs on top of mixture. Bake in 350° oven until egg whites are done. Serve over English muffin and slice of canadian bacon or ham.

Poor man's egg benedict!!

NESTED EGGS

preparation time: 5 minutes   Betsi Dwyer
cooking time: 15 minutes   Malia
serves 6

6 patty shells, top removed,   **1 can soup, cheddar or cream of**
   heated   **mushroom**
6 eggs   **Butter**

Place a pat of butter in each cup of muffin tin. Grease insides of each cup
and crack one egg into each. Bake for 15 minutes in a 350° oven. Cook
soup according to directions, using milk, not water. Place cooked eggs
into heated shells and cover with soup.

CARAVANSARY BREAKFAST

preparation time: 15 minutes   Betsi Dwyer
cooking time: 20 minutes   Malia
serves 6

12 bagels   **¼ lb. your favorite cheese, grated**
12 eggs   **⅓ cup cream**
¼ lb. ham, chopped into small   **Anchovy paste**
   pieces

Split bagels and spread each half with a tsp. of anchovy paste. Broil or
bake in oven. Set aside, but keep warm. Scramble eggs with ham bits
and cream. Cook until eggs are still moist. Cover each bagel half with
some egg mixture, top with cheese and melt.

SCOTCH EGGS

preparation time: 30 minutes   Candy Glover
cooking time: 30 minutes   Anemone
serves 8

8 hard boiled eggs; peeled   **2 eggs, beaten**
1 lb. bulk sausage   **Vegetable oil**
¾ cup seasoned breadcrumbs

Divide bulk sausage into 8 equal parts. Pat sausage around each egg until
eggs are covered evenly. Dip sausage-coated eggs into beaten egg, then
roll in breadcrumbs. In large frying pan, heat oil (about 2″), fry eggs for 5-
6 minutes or until golden brown, turning repeatedly. *Serve hot or cold
with fruit, toast and jam.*

OMELETTE PAYSANNE

preparation time: 20 minutes
cooking time: 15 minutes
serves 5-6

Jennifer Royle
Tri-World

2 Idaho potatoes (¾ lb.)
3 Tblsp. vegetable oil
½ cup very thinly sliced onion
1 cup diced cooked ham
2 Tblsp. parsley
1 Tblsp. tarragon

2 Tblsp. chopped chives
10 eggs
4 Tblsp. butter
Salt
Pepper

Peel potatoes, slice as thin as possible, keep in cold water. Dry. Fry in hot oil. Salt and pepper to taste. Cook about 10 minutes but be sure they don't stick. Brown well, add onions, cook 1 minute. Add ham and 3 Tblsp. butter. Shake skillet and turn so all cook evely. Beat eggs with a whisk. Add spices, pour eggs over ham and potato mixture. Gently stir from bottom allowing eggs to flow underneath. Lift edges of omelette and add 1 Tblsp. butter. Invert on platter.

SOUFFLEÉD OMELETE A LA AVATAR

preparation time: 15-20 minutes
cooking time: 20-30 minutes
serves 4-5

Harriet Beberman
Avatar

4 eggs
1 cup milk
1½ Tblsp. flour
4 tsp. butter

1 cup grated Swiss cheese
1 cup finely chopped ham
Salt and pepper

Separate eggs and in a bowl, mix the yolks, the milk, the flour, pepper and salt using a wooden spoon. Beat the egg whites stiff and gently fold them in the yolk mixture. Resulting batter will be quite liquid, which is the way it is supposed to be. In a 9″ skillet, melt a tsp. of butter. Laddle out enough batter to cover surface of skillet. Reduce heat. Cook batter on one side only. It will resemble a pan cake with a fluffy top, slightly moist. Using a spatula, slide it out of skillet and transfer to overproof dish. Sprinkle with grated cheese. Repeat process, but this time sprinkle with ham. Repeat process alternating ham and cheese. Place fifth pancake moist side down. Place in oven 350° for 20-30 minutes. It will puff up. *Slice into wedges like a cake.*

AVOCADO OMELETTE

preparation time: 5 minutes
cooking time: 5 minutes
serves 1

Harriet Beberman
Avatar

2 eggs
2 Tblsp. water
¼ tsp. salt
1 Tblsp. butter
½ ripe avocado, peeled and sliced

3 thin slices tomato
¼ cup alfalfa sprouts
2 Tblsp. plain yogurt
1 Tblsp. chopped walnuts

Mix eggs, water and salt. Heat butter in 8″ omelet pan until hot. Pour in egg mixture which should set at edges at once. Carefully push cooked eggs up to center tilting pan occasionally so uncooked egg can flow to bottom. While top is moist and creamy looking, arrange one slice of avocado, tomato slices, and sprouts on half of omelet. Fold omelette over filling and turn out onto plate. Top with 2 avocado slices, yogurt and nuts. *Serve with crisp bacon and cranberry muffins.*

SCRAMBLED EGGS, PEPPERS AND TOMATOES

preparation time: 15
cooking time: 15-20 minutes
serves 4-6

Shirley Fortune
Maranatha

2 Medium green peppers, sliced
1 Medium onion, sliced
1 clove garlic, chopped
½ tsp. salt
½ tsp. dried thyme leaves
3 Tblsp. olive oil or margarine
2 medium tomatoes, coarsely
    chopped

8 eggs
½ cup milk
½ cup ¼ inch strips fully cooked
    smoked·ham
1½ tsp. salt
¼ tsp. pepper

Cook and stir green peppers, onion, garlic, ½ teaspoon salt and the thyme in 1 tablespoon of the oil in 10-inch skillet over medium heat until green peppers are crisp-tender, about 8 minutes. Add tomatoes; heat until hot, about 2 minutes. Drain excess liquid from vegetables; place vegetables on platter. Keep warm.
Heat remaining oil in same skillet over medium heat until hot. Mix remaining ingredients; pour into skillet. Cook uncovered over low heat stirring frequently until eggs are thickened throughout but still moist, 3 to 5 minutes. Mound scrambled eggs in center of vegetables. Sprinkle with snipped parsley if desired, and/or paprika.

DUTCH A LA GULLVIVA

preparation time: 10 minutes   *Nancy Thorne*
cooking time: 20 minutes   *Gullviva*
serves 6

½ cup butter   Garnish:
6 eggs   butter
1½ cups milk   powdered sugar
1½ cups flour   lemon juice

Heat oven to 425°. Place butter in 10-inch heavy oven-proof skillet; place in oven to melt butter and heat skillet. In bowl, beat eggs. Add flour and milk and beat just until smooth. Pour batter into hot buttered pan and return to oven. Bake at 425° for 20-25 minutes or until puffed and golden. Cut into wedges. Top with butter, powdered sugar and a squeeze of lemon. Fried sausage and apples also great.

This is a real crowd pleaser and really EEEasy!

PUFF PANCAKE

preparation time: 5 minutes   *Ann Glenn*
cooking time: 20-30 minutes   *Encore*
serves 6-8

1 cup milk   2 Tblsp. butter, melted (or Parkay
3 eggs   liquid)
½ cup flour (whole wheat   ½ tsp. salt
   preferably)   1 Tblsp. sugar (optional)

Grease heavy, straight-sided skillet and place in oven as it preheats to 450°. Combine all ingredients, blend thoroughly and pour into heated skillet. Bake at 450° for 20 minutes. Then reduce temperature to 375° for another 5-10 minutes or until pancake is beautifully brown. The pancake will have climbed up the sides to form a puffy panshape. Serve from the skillet, or slip onto a serving platter before cutting into sections. Let each person layer his pancake with warmed apple slices and cold cottage cheese (or any combination of fresh fruit, yogurt, applesauce, etc.)
The contrast between hot and cold layers is surprising and delightful.

• *After using hand can opener, roll a piece of folded paper towel through the can opener. This will stop rust as well as clean food from the opener.*

DANISH AEBLESKIVERS

preparation time: 30 minutes
cooking time: 2-3 minutes each
serves 8

Roy Olsen
Kingsport

6 eggs, separated
2 cups milk
3 cups flour
1 Tblsp. baking powder
1 Tblsp. cardamom
20 oz. can apple pie filling, season
   with extra cinnamon

3 Tblsp. sugar
¾ tsp. salt
Juice and grated rind of 1 lemon
Apple slices or a mashed banana

Beat egg yolks and sugar until light. Add lemon juice, lemon rind, salt, baking powder, cardamom, sifted flour and milk. Fold in egg whites that have been beaten until fluffy. Grease aebleskiver pan and place over medium heat. Fill cups ¾ full and place an apple slice or mashed bananas in center of each. When bottom side is light brown, turn with a fork. Excellent served with maple syrup, lemon sauce or devonshire cream (see Index). These are something like a donut with a filling. They can also be made like pancakes on a griddle.
This is an Olsen family recipe coming from my grandmother, Catherine Olsen.

LEMON SAUCE:

1 cup sugar
1 egg
¼ lb. butter

3 Tblsp. lemon juice
3 Tblsp. lemon rind

Heat all ingredients in saucepan over low heat, stirring until it boils and thickens. *Serve warm.*

POTATO PANCAKES

preparation time: 10 minutes
cooking time: 15 minutes
serves 4-6

Cheryl Fleming
Silver Queen of Aspen

¼ cup milk
2 eggs
3 cups diced raw potatoes
1 small onion, quartered

4 Tblsp. flour
1¾ tsp. salt
¼ heaping tsp. baking powder

Put all ingredients in blender, blend about 10 seconds (do not over blend or potatoes will liquify). Use Crisco for a crispy and not greasy pancake. Garnish with applesauce and sour cream on the side.

ORANGE YOGURT PANCAKES WITH STRAWBERRIES

preparation time: 15 minutes   *Jean Thayer*
cooking time: 20 minutes   *Paradise*
serves 4

1 pkg. frozen strawberries,
   defrosted
3 Tblsp. sugar
Grated rind of 1 orange
⅓ cup orange juice
¾ cup plain yogurt + ¼ cup for
   garnish

1 egg
2 Tblsp. melted butter
1 cup flour, sifted
1 tsp. baking soda
½ tsp. double-acting baking
   powder
¼ tsp. salt

In a small bowl combine grated rind, juice and sugar stirring to dissolve sugar. Add yogurt, egg, and melted butter, mixing well. In another bowl, sift together flour, baking soda, baking powder and salt. Add yogurt mix to flour for a thick batter. Heat griddle; brush with butter and spoon out batter to form 3″ round pancakes. Cook cakes till golden and turn. Remove to 200° oven and continue. *To serve top with strawberries and a dollop of yogurt.*

PINA COLADA PANCAKES

preparation time: 15 minutes   *Kandy Popkes*
cooking time: 20 minutes   *Wind Song Oregon*
serves 4-6

2 cups Bisquick
2 eggs
¾ cup milk (approximately)
4 Tblsp. Coco Lopez (cream of
   coconut)
2 tsps. vanilla or 2 Tblsp. Kahlua

3 Tblsp. coconut
½ cup diced pineapple, fresh or
   canned
1 diced banana
1 orange sliced for garnish

Make pancakes with Bisquick, eggs, Coco Lopez, Kahlua and enough milk to make a firm consistency. Heat and butter griddle. Spoon enough batter for 4-inch pancakes. Turn and remove when both sides are browned. Place on hot platter and keep warm. Continue process until all batter is used. Sprinkle with diced fruit and coconut. *Serve with warm syrup. Garnish with orange slices.*

Syrup: equal amounts of Coco Lopez and Maple Syrup

ORANGE PANCAKES

preparation time; 10 minutes   *Katie MacDonald*
cooking time: 10 minutes   *Oklahoma Crude II*
serves 6

Pancake mix (water base)   **½ cup wheat germ (optional — for
Orange juice**   **more hearty pancakes)**
Brown sugar

Prepare mix as directed substituting orange juice for water. Sweeten with brown sugar to taste. Top with a pad of butter and sifted confectionery sugar or orange sauce.

V.I. FRENCH TOAST

preparation time: 10 minutes   *Katie MacDonald*
cooking time: 10 minutes   *Oklahoma Crude II*
serves 4

6 eggs   **8 slices cinnamon raisin bread**
¼ cup cream   **1 tsp. nutmeg**
¼ cup Cruzan rum   **1 Tblsp. brown sugar**

Mix all ingredients except bread. Beat well. Soak bread about 1 minute before cooking. Cook over medium heat until golden brown and puffy. Top with a pat of butter and sifted confectionery sugar. (No syrup necessary.)

GAY'S BERRY PANCAKE

preparation time: 25 minutes   *Gay Thompson*
cooking time: 45 minutes   *Satori*
serves 8

6 eggs   **½ cup butter**
1½ cups milk   **Powdered sugar**
1½ cups flour
**1 can blackberries or red
    raspberries**

Slightly beat 6 eggs. Add 1½ cups milk. Slowly add 1½ cup flour beating with wire whisk or use your blender. Melt butter in 9x13 oven-proof pan. Drain and rinse berries. Add to flour mixture. Pour in pan containing butter. (Butter will rise to top.) Bake at 350° about 45 minutes. It will rise to mountains and valleys, pull away from edges and be brown on the "hills" and edges. *Falls quickly so serve immediately with cooked bacon and champagne. Sprinkle powdered sugar on top of the pancake.*

SKIPS TIPSY TOAST

preparation time: 30 minutes   *Paul Taylor*
cooking time: 10 minutes   *Viking Maiden*
serves 12

12 eggs
1 cup milk
1 cup orange juice concentrate

1 jigger Grand Marnier
Liquid margarine
2-3 loaves French Bread

Combine all ingredients in large bowl, except bread. Slice French bread, soak bread in mix for at least 15 minutes and fry in liquid margarine until crisp and golden.

PERO'S RUM BANANA PANCAKES

preparation time: 15 minutes   *Bob Robinson*
cooking time: 20 minutes   *Vanity*
serves 6-8

2 cups Aunt Jemina Pancake Mix
1 cup milk
1 egg
1 Tblsp. vegetable oil

1 banana, riper the better
⅓ cup rum
Syrup — see below

Combine all ingredients in a blender and blend. Preheat skillet and spoon enough batter for a 3-inch wide pancake. Cook until golden on each side. Remove and place on a hot platter and keep warm in the oven. Continue process until all batter is used. Garnish with orange twists. *Serve with Syrup: — Heat strawberry preserves and butter, stir and add a little rum (or as much as you wish). A wonderful start to the day!!*

GOOD MORNING APPLE KUCHEN

preparation time: 20 minutes   *Lori Moreau*
cooking time: 35 minutes   *Alberta Rose*
serves 6

½ cup butter
1 pkg. Yellow Cake Mix
½ cup flaked coconut
2½ cups apples sliced

½ cup sugar
1 tsp. cinnamon
1 cup sour cream
2 egg yolks

Cut butter into cake mix until crumbly. Mix in coconut. Press lightly into ungreased 13 x 9 x 2 pan (build up edges). Bake 10 minutes at 350°. Arrange apples on warm crust. Sprinkle with sugar and cinnamon. Blend sour cream and egg yolks. Drizzle over apples (will not completely cover). Bake 25 minutes at 350°.

ANN'S GOOD GRANOLA

preparation time: 10 *minutes*   *Ann Glenn*
cooking time: 25-30 *minutes*   *Encore*
serves 12

4 cups rolled oats
1 cup almonds, or other nuts,
   chopped
½ cup each: sunflower nuts,
   sesame seeds, wheat germ
½ cup each: raisins, coconut,
   apricot

¼ cup dry milk powder
½ cup each: peanut oil and honey
1 Tblsp. vanilla essence
1 tsp each: almond, lemon, and
   maple
1 tsp. cinnamon

Mix all dry ingredients together in a large shallow pan, stir well. In a small saucepan heat and stir oil, honey and cinnamon until blended. Mixture will foam and rise as it comes to the boil. Remove from the fire. Add the essences, stir thoroughly. Drizzle the oil mixture over the dry ingredients and mix to coat evenly. Bake in preheated 300° oven about 30 minutes stirring thoroughly each 10 minutes. Cool. Store in airtight, cool, dry place. *Served with Cinnamon Bran Muffins and Apple Juice makes this our double-barrelled health food breakfast.*

FRUIT GRANOLA

preparation time: 10 *minutes*   *Kathleen McNally*
cooking time: 15 *minutes*   *Orka*
Makes about 10 *cups*

4 cups rolled oats
1¾ cups rolled wheat flakes*
1 cup sunflower seeds
1 cup chopped nuts (almonds,
   walnuts, pecans, or whatever)
⅓ cup toasted sesame seeds

⅓ cup sunflower oil
¼ cup honey
½ cup raisins
½ cup chopped dried apricots
½ cup chopped dried peaches
½ cup chopped dried apples

Preheat oven to 400°. Blend oats, wheat flakes, sunflower seeds, nuts, seasame seeds, oil and honey in large bowl. Transfer to large roasting pan spreading evenly. Bake until lightly browned at edges, stirring occasionally about 15 minutes. Let cool completely. Stir in fruit. Store in airtight container.

*available in natural food stores

FRUIT GRANOLA GRATIN

preparation time: 5 minutes
cooking time: 25 minutes
serves 8-10

Kathleen McNally
Orka

4 cups Fruit Granola (see recipe)
2 cups milk (long life)
2 cups water
¼ cup firmly packed brown sugar

⅛ tsp. salt
Whipping cream (garnish)
Brown sugar (garnish)

Preheat oven to 375°. Mix all ingredients except garnish in large bowl. Turn into shallow 14 to 16-inch gratin pan or baking dish. Bake until all liquid is absorbed and top is browned, about 25 minutes. *Serve gratin warm with cream and brown sugar.*

JUST PEACHY

preparation time: 10 minutes
cooking time: 15-20 minutes
serves 4

Gay Thompson
Satori

21 oz. can peach halves
Granola

Brown Sugar
Butter

In a deep dish pan, place 2 peach halves per person. Fill center cavity with granola. Sprinkle with brown sugar. Dab a pat of butter on each. Pour peach juice in bottom of pan. Bake for 15-20 minutes at 350°, basting once with juice. *Serve with English muffins and crisp bacon. The whole flavor is delicious!*

BACON CASSEROLE

preparation time: 15 minutes
marinating time: overnight
cooking time: 45 minutes
serves 8

Debbie Olsen
Kingsport

10 slices white bread, buttered
   and cubed
1 lb. grated cheddar cheese
1 qt. milk
6 eggs

2 tsp. dry mustard
2 tsp. salt
½ lb. bacon, fried crisp and
   crumbled

Put bread cubes in 9x13 pan, toss and cover with cheese. Combine milk, eggs, mustard and salt in a blender, then pour over cheese. Top with bacon and refrigerate overnight. Bake 45 minutes at 350°. This recipe comes from my Mother, Jeanne Patterson.

FRUIT CREPES

preparation time: 10 minutes
cooking time: 10 minutes
serves 6

K.S. Strassel
Great Escape

6 (6-8 inch) crepes (make ahead and freeze)
2 apples, peeled, cored, diced
1 cup strawberries, sliced
1 cup peaches, sliced

2 bananas, sliced
1 cup vanilla yogurt
½ cup granola
½ cup raisins
Caramel Nut Sauce

Prepare (or thaw) crepes — set aside. Prepare Caramel Nut Sauce — set aside to cool. Mix fruits, yogurt, granola and raisins. Roll fruit mixture in crepes, warm under broiler for 1 minute. *Top with Caramel Nut Sauce and serve with bacon. A great Sunday breakfast/brunch.*
CARMEL NUT SAUCE:

4 tsp. butter
½ cup slivered almonds or chopped pecans

1 cup firmly packed brown sugar
1 cup whipping cream

Melt butter in frying pan. Add nuts and cook, stirring frequently until lightly toasted. Stir in sugar and whipping cream; cook, stirring until sauce boils and sugar dissolves. Remove from heat and cool. *Serve at room temperature. Best served same day as made. This recipe is from Ann Glenn on Encore.*

WELSH RAREBIT

preparation time: 15 minutes
cooking time: 15 minutes
serves 7-10

Candy Glover
Anemone

¼ cup butter
¼ cup flour
½ tsp. salt
½ tsp. dry mustard
½ tsp. Worcestershire sauce

¾ cup milk
¾ cup beer
2 cups cheddar cheese, grated
10 slices Canadian bacon
10 slices tomato
10 English muffin halves

Heat butter in saucepan until melted. Stir in flour, salt, dry mustard, and Worcestershire sauce. Cook over low heat until smooth stirring constantly. Stir in milk and beer. Boil and stir for about 1 minute. Stir in cheese. Reduce heat and stir constantly until melted and smooth. Toast English muffins until lightly browned. Heat Canadian bacon in microwave or under broiler. Place one slice bacon, one slice tomato on each muffin half. Arrange evenly on cookie sheet. Top each with a spoonful of sauce. Broil or bake in oven until bubbly and light brown. Garnish with parsley.

APPLE SAUSAGE TOSS

preparation time: 10 minutes   *Jean Thayer*
cooking time: 20 minutes   *Paradise*
serves 6

½ lb. pork sausage links   2 Tblsp. butter
3 tart apples, peeled, cored, cut   1 Tblsp. sugar, or to taste
   into ⅛'s   Cinnamon, to taste

Cook sausage according to directions. Remove from pan, drain and cut into bite-sized pieces. In medium frying pan, melt butter, add slices of apple cooking till tender. Add sugar, tossing to coat. Add sausage and cook till heated through. Sprinkle with cinnamon to taste. *Serve with Corn Fritters.*

GLAZED CANADIAN BACON

preparation time: 5 minutes   *Kathleen McNally*
cooking time: 1 hour   *Orka*
serves 8-10

1 (3-5 lb.) whole Canadian bacon   ½ cup maple syrup
½ to 1 cup apple cider (or other   2 tsp. dry mustard
   unsweetened fruit juice) beer   Whole cloves
   or ginger ale

Preheat oven to 325°. Set bacon in baking pan and pour juice over. Bake 15 minutes per pound basting every 15 minutes with juices. Whisk maple syrup and mustard in small bowl. About 15 minutes before end of baking time, score outside of bacon diagonally to form diamonds. Stud decoratively with cloves. Brush with syrup mixture and continue baking until brown and well-glazed, basting once or twice. Transfer to heated platter. *Slice thinly to serve.*

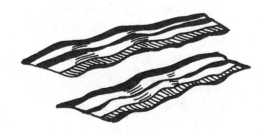

• *Breakfast bacon fried in oven will not curl.*

CREAMED MUSHROOMS ON TOAST

preparation time: 10 minutes
cooking time: 10 minutes
serves 6

K. A. Strassel
Great Escape

2 cups fresh mushrooms, sliced
2 Tblsp. flour
2 Tblsp. butter
2 cups milk
Salt
Pepper

Thyme
Tarragon
2 Tblsp. Worcestershire
¼ cup cream cheese
Toast

Melt butter, whisk in flour to make a paste. Add milk slowly and heat to boiling. Set aside on warm. Sauté mushrooms for 2 minutes, add to cream sauce. Add cream cheese, Worcestershire and seasonings to taste. Heat and stir until smooth. *Serve on toast, garnish with parsley. Fresh fruit compote is a good accompaniment.*

POPOVERS (YORKSHIRE PUDDING)

preparation time: 30 minutes
rising time: 1 hour
cooking time: 30 minutes
serves 6

Terry Boudreau
High Barbaree

⅞ cup flour
½ tsp. salt
½ cup milk

2 eggs
½ cup water

The ingredients must be at room temperature when they are mixed or they will not puff.

Sift into a bowl the flour and salt. Make a well in the center and put in ½ cup milk. Stir in the milk. Beat the eggs until fluffy and beat them into the batter. Add ½ cup water. Beat the batter well until large bubbles rise to the surface. Allow the batter to stand covered in the refrigerator for 1 hour. When ready to bake, have ready a hot muffin tin containing about ¼ inch hot shortening or butter. Pour in the batter about ⅜ inch high. Bake the puddings in a hot 400° oven for twenty minutes. Reduce heat to 350° and cook 10 to 15 minutes more. *Serve at once.*

RYE HEART TOASTS

preparation time: 10 minutes   *Kathleen McNally*
cooking time: 15 minutes   *Orka*

**1 lb. loaf of rye bread, very thinly
sliced**

Pre-heat oven to 375° F. Using 2″ heart-shaped cookie cutter, cut out as many hearts as desired from bread. (Reserve trimmings to make croutons or crumbs for another use.) Arrange hearts in single layer on ungreased baking sheets. Bake, turning once until crisp and lightly colored, about 4-6 minutes per side. Transfer to rack to cool.

GRUYERE GARLIC TOAST

preparation time: 10 minutes   *Kathleen McNally*
cooking time: 10 minutes   *Orka*

**1 loaf French or Italian bread
1-2 garlic cloves, halved
3 Tblsp. olive oil**

**½ lb. sliced Gruyere cheese (for
stronger flavor use ⅓ cup
freshly grated parmesan
cheese instead. Sprinkle
before second baking.)**

Preheat oven to 425°. Line baking sheets with foil. Arrange sliced bread on foil and toast. Remove from oven and rub surfaces with cut side of garlic. Drizzle with olive oil; top with Gruyere. Bake until cheese is lightly browned — 5-7 minutes. *Serve immediately.*

GOUGERES (CHEESE PUFFS)

preparation time: 20 minutes   *Louella Holston*
cooking time: 35 minutes   *Eudroma*
makes 16-20

**1 cup water
½ cup (1 stick) butter
½ tsp. salt
1 cup flour
4 eggs**

**Dash of cayenne pepper
Dash of nutmeg
1-1½ cups Gruyere cheese,
shredded**

Place water, butter, and spices in sauce pan until all boils. Add flour all at once. Beat with wooden spoon until mixture pulls away from sides of pan. Remove from heat and beat until slightly cooled — 30 seconds.

Add eggs one at a time beating one minute after each egg. Stir in cheese until well blended. Drop by tablespoonfuls onto a greased cookie sheet at least 2 inches apart. Bake at 375° until well browned, about 35 minutes. Cut slit in each to let steam escape.

GOUGÈRE

preparation time: 15 minutes
cooking time: 40 minutes
serves 6

Jean Thayer
Paradise

1 cup water	1 tsp. Dijon mustard
6 Tblsp. butter	1 Cup flour
1½ tsp. salt	4 eggs
Fresh ground pepper	1 cup grated Swiss cheese

Heat water, butter, salt, pepper and mustard until it boils rapidly. Over low heat, add flour all at once. Stir until mixture forms a ball in the middle of the pan. Beat in eggs, one at a time until dough is smooth. Stir in all but 2 Tblsp. of the cheese. Place dough on a greased cookie sheet in the shape of a ring dropping by an ice cream scoop or large spoon. Sprinkle top with remaining cheese.

Bake at 425° for 35-45 minutes or 'til puffed and golden. Serve hot with butter.

This can be served as an hors d'oeuvre — but scoops of dough should be smaller or tablespoon size.

BISCUIT MIX

preparation time: 15 minutes
storing time: indefinitely
makes 11 cups

Terry Boudreau
High Barbaree

8 cups sifted all-purpose flour	1 cup skim milk powder
5 Tblsp. baking powder	1½ cups shortening
2 tsp. salt	

Sift flour, baking powder and salt. Add skim milk powder and stir until thoroughly mixed. Cut in shortening until mixture resembles fine bread crumbs. Store in a covered container in refrigerator or other cool place.

TEA BISCUITS

preparation time: 15 minutes
cooking time: 15 minutes
makes 16 biscuits

2½ cups firmly packed Biscuit Mix   ⅔ cups water
　　(recipe above)

Stir Mix and water to make a soft dough. Turn out on floured board and knead gently. Roll to about ¾ inch thickness, cut into 2-inch rounds. Bake 15 minutes at 425°. Makes 16 biscuits.

QUICK WHOLE WHEAT LOAVES

preparation time: 20 minutes
rising time: 45-60 minutes
cooking time: 30-40 minutes
makes: 2 loaves (or 16 rolls)

Jan Robinson
Vanity

2 cups whole wheat flour
2 cups all-purpose flour
2 tsp. salt
2 tsp. sugar
1 Tblsp. lard

1 pkg. active dry yeast
1¼ cups warm water
3-4 Tblsp. cracked wheat or
   crushed cornflakes

Sift the two flours, salt and sugar into a bowl. Cut up lard and rub into flour. Blend yeast with the warm water until dissolved and frothy. Make well in flour mixture and pour in yeast liquid. Mix, beating until it leaves the sides of the bowl clean (if necessary add a little more flour). Divide dough into 2; shape each portion to half, fill a greased 1-pound loaf pan. Brush top with lightly salted water. Sprinkle with cracked wheat or crushed cornflakes. Cover with plastic wrap or a towel and leave in warm place to double. Bake the loaves in the centre of a preheated oven at 450° for about 40 minutes. Test for doneness by tapping the bottom of the loaves, if they sound hollow, they're done. Cool loaves on wire rack.

A Flower of a thought: A clay flowerpot, 4-5 inches wide will hold the dough of the above recipe. Butter the clay flowerpot thoroughly on the inside and bake empty in a hot oven a couple of times, allowing it to cool. This will seal the inner surface and prevent the dough from sticking.

BEER BREAD

preparation time: 5 minutes
cooking time: 1 hour
cooling time: 30 minutes
makes: 2 loaves

Debbie Olsen
Kingsport

6 cups self-rising flour
4 Tblsp. sugar
2 Tblsp. dillweed

2 cans beer
caraway seeds

Mix all ingredients together and pour into two greased loaf pans. Sprinkle caraway seeds on top and bake for one hour at 375°. Cool on rack at least ½ hour before slicing.

EASIEST BEER BREAD

preparation time: 10 minutes
cooking time: 45 minutes
Makes 1 loaf

Lori Moreau
Alberta Rose

3 cups self-rising flour
3 Tblsp. sugar

1-12 oz. can beer (room
    temperature)

Preheat oven to 350°. Combine ingredients, mixing only until all are moistened. Pour into greased bread pan and bake approximately 45 minutes. Does not brown as much as regular bread.

HOMEMADE WHITE BREAD BRAID

preparation time: 30 minutes
rising time: 1¼ hours
cooking time: 30-40 minutes
makes: 2 braids

D.J. Parker
Trekker

¼ cup lukewarm water
1½ package dry yeast
¼ cup raw honey
2 cups lukewarm milk

4 Tblsp. butter, melted
2 tsp. salt
6 to 6½ cup all purpose flour,
    divided

In large mixing bowl mix water, yeast, and honey until yeast dissolves. Stir in milk, butter and salt until well blended. Using sturdy wooden spoon mix in one cup flour until smooth. Add two cups of flour and beat vigorously. Slowly add about two more cups of flour. Sprinkle with flour and gently knead in bowl adding flour if necessary. Oil sides and bottom of bowl and let rise until double (about 45 minutes). Punch down, divide into six parts. Gently knead and roll into lengths about 12-18 inches. Press ends of three lengths together and proceed to braid pressing other ends together and turn under. Braid remainder and place both braids on lightly greased baking sheet. Let rise until almost double (about 30 minutes). Brush tops with water. Bake in 350° oven approximately 30 minutes until brown and has a hollow sound when tapped. *Serve warm.*

• *Dry stale bread in a slow oven. Put in a freezer plastic or paper bag and crush to make bread crumbs.*

CORN FRITTERS

preparation time: 10 minutes
cooking time: 10 minutes
serves 6

Jean Thayer
Paradise

2 eggs
1 cup flour
2 tsp. baking powder
1 small 8 oz. can cream corn

¼ cup sugar
Confectionery sugar
Vegetable oil for frying

Mix flour, baking powder, eggs, corn and sugar together. In medium-size saucepan, pour vegetable oil to a depth of 3″ and heat till very hot. Drop corn mixture by teaspoonfuls into hot oil. Cook till golden and reserve in 200° oven on a paper towel-lined baking sheet. Continue with remaining batter. *To serve, dust gently with confectionary sugar. These fritters are great served with Apple Sausage Toss.*

CUTIE'S FRIED CORN BREAD

preparation time: 5 minutes
cooking time: 12-15 minutes
serves 6

Jan Robinson
Vanity

⅔ cup self-rising cornmeal
2 Tblsp. plain flour
1 egg

8 Tblsp. bacon drippings
Water

Preheat electric skillet to 380°. Add bacon drippings. Combine the flours and mix in the egg and enough cold water to make a thick batter. Place a big tablespoon of batter in the skillet, repeat, do not let sides touch. Brown on each side, remove and eat . . . delicious. Teaspoonful size makes great hors d'oeuvres. NOTE: Preheat skillet before adding water to flour.

CORN BREAD

preparation time: 10 minutes
cooking time: 20 minutes
serves 8

Katie MacDonald
Oklahoma Crude II

1 cup flour (whole wheat for a
    more grainy bread)
1 cup cornmeal
½ tsp. soda
2 tsp. baking powder

⅓ cup sugar or ¼ cup honey
1 cup milk, water or buttermilk
1 egg (beaten)
¼ cup melted butter

Mix all ingredients. Spread in greased 8 x 8 pan and bake at 400° for 20-25 minutes.

COCONUT BREAD

preparation time: 15 minutes
cooking time: 45 minutes
serves 6

Katie MacDonald
Oklahoma Crude

2 cups flour
½ tsp. salt
1 tsp. nutmeg
1 tsp. cinnamon
2 tsp. baking powder
½ tsp. baking soda

½ cup brown sugar
¼ cup melted butter
2 cups coconut
2 eggs
1 tsp. vanilla

Mix all ingredients thoroughly. Spread in greased and floured 8 x 8 pan. Bake at 350° for approximately 45 minutes or until bread edges away from pan. It will remain a bit moist inside.

PAULA'S WEST INDIAN COCONUT BREAD

preparation time: 15 minutes
cooking time: 1 hour
serves 10-12

Paula Taylor
Viking Maiden

7 oz. shredded coconut
4 cups flour
2 cups sugar
1 tsp. cinnamon
1 tsp. nutmeg
2 tsp. vanilla

2 tsp. baking powder
1 cup milk
½ cup Coco Lopez
4 Tblsp. sour cream
1 tsp. salt
4 beaten eggs

Mix all ingredients — mixture will be thick. Separate into two 5 x 9 loaf pans and cook at 350° for an hour or until done. *Serve warm with butter as an accompaniment to to Chowder or is absolutely super with cheese board. Great cocktail snack spread with cream cheese.*

LEMON NUT BREAD

preparation time: 15 minutes
cooking time: 50 minutes
makes 1 loaf

Cheryl Fleming
Silver Queen of Aspen

3 oz. sugar
¼ cup butter
2 eggs
2 tsp. lemon juice
1 tsp. vanilla

1 cup yogurt
2 cups all-purpose flour
1 tsp. baking soda
1 tsp. baking powder
¼ tsp. salt

Cream butter and sugar over a pan of water on the stove. Sift together dry ingredients. Add together all ingredients and mix well. Grease 9 x 5 x 3 loaf pan. Bake at 350° for 50 minutes.

ZUCCHINI-CARROT NUT BREAD

preparation time: 15 minutes
cooking time: 40 minutes

Kathleen Leddy
Carriba

2½ cups flour (half white/half
    whole wheat)
¼ cup powdered milk
¼ cup wheat germ
2 tsp. baking soda
½ tsp. baking powder
1 cup sugar (half brown/half
    white)

3 tsp. cinnamon
½ tsp. nutmeg
1 cup oil
3 eggs, beaten
3 tsp. vanilla
1 cup chopped nuts
3 cups mixed, grated carrots and
    zucchini

Combine all dry ingredients in large bowl. Mix eggs, vanilla, and oil together then stir into dry ingredients. Add grated zucchini, carrots and nuts. Bake in well greased and floured loaf pans or makes delicious muffins. Bake in 350° oven. *Can be served for breakfast, snacks or dessert with cream cheese.*

ZUCCHINI NUT BREAD

preparation time: 20 minutes
cooking time: 1 hr. 15 minutes
makes 1 loaf

Chris Kling
Sunrise

2 cups flour
½ tsp. baking soda
½ tsp. baking powder
½ tsp. salt
2 tsp. cinnamon
1½ cup grated zucchini

2 eggs, beaten
1½ cups sugar
1 tsp. vanilla
⅓ cup oil
½ cup chopped nuts

Preheat oven to 375°. Combine all ingredients and pour into a greased loaf pan. Bake for 1 hr. 15 minutes. Cool before slicing.

BANANA NUT BREAD

preparation time: 20 minutes
cooking time: 50-60 minutes
serves 6-8

Trish Penn
Daddy Warbucks

½ cup margarine
¾ cup sugar
1 egg
1 Tblsp. lemon juice
2 cups flour

1 tsp. soda
½ tsp. salt
1 cup mashed bananas
1 cup chopped walnuts

Cream sugar and margarine. Add egg and beat thoroughly. Add sifted dry ingredients. Mix well. Stir in mashed bananas, lemon juice and walnuts. Fill one 9 x 5 x 3 loaf pan — ungreased. Bake at 350°.

BANANA BREAD

preparation time: 15 minutes
cooking time: 1 hour
makes 1 loaf

Louella Holston
Eudroma

2 cups mashed very ripe bananas
8 oz. butter — creamed with 1½ cups sugar
1 Tblsp. baking powder
4 eggs

vanilla
2 cups flour
½ tsp. salt
nuts

Mix all ingredients in a bowl. Pour into a tube pan. Bake at 350° about 1 hr.

MY BEST GINGERBREAD

preparation time: 20 minutes
cooking time: 45 minutes

Terry Boudreau
High Barbaree

½ cup butter or shortening
½ cup granulated sugar
1 egg, beaten
2½ cups flour, sifted
1½ tsp. baking soda
1 tsp. cinnamon

1 tsp. ginger
½ tsp. cloves
½ tsp. salt
1 cup Brer Rabbit Molasses
1 cup water (hot)

Cream shortening and sugar. Add beaten egg. Measure and sift ingredients. Combine molasses and hot water. Add dry ingredients alternately with liquid, a small amount at a time, and beat after each addition until smooth. Bake in greased and floured pan 9x9x2 in moderate oven — 350° for 45 minutes.

ISLAND COFFEE CAKE

preparation time: 5 minutes　　　　　　　　　　*Ann Glenn*
cooking time: 25 minutes　　　　　　　　　　　　　　*Encore*
serves 8

¼ cup melted butter or Parkay
1 cup yogurt, lemon or pineapple
1 egg
2 cups Bisquick
½ cup sugar

¾ cup shredded coconut
½ cup crushed pineapple, well
　drained
Extra coconut for the top

Combine butter, yogurt and egg. Stir in Bisquick and sugar, coconut and pineapple, mixing well. Spread in greased 9x13 pan and sprinkle with extra coconut. Bake at 350° about 25 minutes, or until lightly browned. *Serve with Sherried Ham and Eggs.*

PINEAPPLE CRUNCH COFFEE CAKE

preparation time: 20 minutes　　　　　　　*Jeannie Drinkwine*
cooking time: 35 minutes　　　　　　　　　　　　　　*Vanda*
serves 6

2 cups flour
1½ cups sugar
1 tsp. soda
½ tsp. salt
¼ cup packed brown sugar
1 cup chopped walnuts
1 (1 lb. 4 oz.) can crushed
　pineapple

ICING:
⅔ cup sugar
¼ cup milk
½ cup margarine or butter

Combine flour, sugar, soda, salt, brown sugar and pineapple with juice and nuts. Mix well. Pour into ungreased 9x13 loaf pan. Bake at 350° for 35 minutes.

Icing: Combine the above in a small saucepan and boil 2 minutes. Pour over warm cake. Sprinkle with coconut if desired.

CREPES

preparation time: 20 minutes
Makes 16-18 Crepes

Candy Glover
Anemone

1 cup flour
1½ cups milk
2 eggs

1 Tblsp. cooking oil
¼ tsp. salt

In a bowl, mix flour, milk, eggs, cooking oil and salt. Whip with wire whisk until smooth. In crepe pan or 6-9 inch round skillet put drop of oil and wipe evenly with a paper towel. Pour in 2 Tblsp. batter. Lift and tilt skillet until batter covers bottom of pan. Brown on one side only. Slip crepe onto empty plate and repeat. I generally make these up ahead of charter and freeze them. To do this, put oiled waxed paper between each crepe, stack them and wrap the whole pile in foil. To reheat crepes, wrap 3 or 4 in foil and heat through in the oven at 300° F., or they reheat great in a microwave (but not in foil).

For dessert crepes, sift 1 Tblsp. sugar with the flour and salt and add 1 tsp. of vanilla extract to the batter. Liqueur can be substituted for the vanilla extract.

BLINTZ BUBBLE RING

preparation time: 15 minutes
cooking time: 20 minutes
serves 8-10

Ann Glenn
Encore

2 pkgs. refrigerated biscuits
2 pkgs. cream cheese (3 oz. each)
½ cup sugar
1 tsp. cinnamon

3 Tblsp. butter, melted (liquid Parkay)
½ cup pecans or walnuts, chopped

Preheat oven to 375°. Cut one pkg. of cream cheese into 10 cubes. Open one roll of biscuits; roll or pat each biscuit to 3" diameter. Combine sugar and cinnamon. Place 1 tsp. sugar mix and 1 cheese cube on each biscuit. Fold over and pinch to seal edges. Pour butter into bottom of 5 cup ring mold. Sprinkle ½ of nuts, then ½ of remaining sugar mix into mold. Place rolls in a layer around the ring. Sprinkle remaining nuts and ½ of sugar mix over rolls. Repeat process of roll, fill, and fold, placing a second layer in the ring mold. Bake at 375° about 20 minutes. Cool 5 minutes, then invert on platter to serve.

CRANBERRY MUFFINS

preparation time: 20 minutes　　　　　　　　*Kathleen McNally*
cooking time: 25 minutes　　　　　　　　　　　　　*Orka*
makes 24 muffins

2 cups unbleached all-purpose
　　flour
1 cup whole wheat flour
4 tsps. baking powder
1 tsp. salt
½ tsp. baking soda
⅔ cup butter, room temperature
1 cup sugar

2 eggs
1 cup milk
2 cups cranberries, halved and
　　tossed in ¼ cup all-purpose
　　flour
4-6 Tblsp. sugar
Grated peel of 2 medium lemons

Sift flours, baking powder, salt and baking soda; set aside. Cream butter in large bowl. Gradually add 1 cup sugar, beating until smooth. Beat in eggs. Blend milk into batter alternately with sifted ingredients. Stir in berries. Spoon into prepared muffin pans. Combine 4-6 Tblsp. sugar with lemon peel in small bowl rubbing between fingertips until sugar is moistened. Sprinkle evenly over batter. Bake until muffins are light golden and tester inserted comes out clean. About 25 minutes. *Serve warm or at room temperature.*

BANANA-RUM MUFFINS

preparation time: 10 minutes　　　　　　　　*Jennifer Royle*
cooking time: 15 minutes　　　　　　　　　　　*Tri-World*
serves 8

3 cups Bisquick
1 cup mashed bananas (about 2
　　medium)
½ cup brown sugar
¼ cup rum
1 egg

½ cup chopped nuts
Glaze:
¾ cup powdered sugar with 1
　　Tblsp. plus 1 tsp. water until
　　smooth

Preheat oven to 400°. Mix bisquick, bananas, brown sugar, rum and egg (batter will be lumpy). Fill muffin tins about ½ to ¾ full. Bake about 15 minutes or until light brown. Remove from pan, cool 5 minutes. Dip each muffin into glaze and then chopped nuts.

• *Soften brown sugar by placing in warm oven. When removed from oven, refrigerate.*

BLUEBERRY MUFFINS

preparation time: 10 minutes   *Jessica Adam*
cooking time: 20-25 minutes   *Nani Ola*
serves 6-8

1¾ cups sifted flour   ¾ cup milk
¼ cup sugar   ⅓ cup salad oil
2½ tsp. baking powder   1 cup fresh or well-drained frozen
¾ tsp. salt   blueberries
1 well beaten egg

Sift dry ingredients into bowl; make well in center. Combine egg, milk
and oil. Add all at once to dry ingredients. Stir quickly just until dry ingredients are moistened. Add blueberries until greased muffin tins are ⅔
full. Bake at 400° for 20-25 minutes. Makes 10.

MARMALADE MUFFINS

preparation time: 10 minutes   *Jennifer Royle*
cooking time: 15 minutes   *Tri-World*
serves 3-4

1 egg, beaten   1 Tblsp. salad oil
¼ cup orange juice   1 cup Bisquick
2 Tblsp. sugar   ¼ cup orange marmalade

Preheat oven to 400°. Combine egg, orange juice, sugar and oil. Add Bisquick and beat 30 seconds. Stir in marmalade. Fill greased tins ⅔ full
and bake 15-20 minutes. Makes 6-8 large muffins.

MAGIC MUFFINS

preparation time: 15 minutes   *Debbie Olsen*
cooking time: 10 minutes   *Kingsport*
serves 8

2 tubes Pillsbury Crescent Rolls   ½ cup sugar
¼ lb. butter, melted   ½ cup cinnamon
1 pkg. large white marshmallows

Dip marshmallows in melted butter, then into cinnamon-sugar mixture,
then back into melted butter. Wrap crescent roll around marshmallow and
pinch edges together. Bake in muffin tray at 350° F. approximately 10
minutes or until golden brown. This recipe came from Sean Frost, our
charter guest.

REFRIGERATED BRAN MUFFINS

preparation time: 20 minutes
cooking time: 20-25 minutes

Launa Cable
Antiquity

2 cups 100% Bran
2 cups boiling water
3 cups sugar
1 cup Crisco oil or shortening
4 eggs

1 qt. Buttermilk
4 cups All Bran
5 cups flour
1 tsp. salt
5 tsp. baking soda

Soak 100% Bran in boiling water while preparing the rest. Cream sugar, shortening, eggs, add buttermilk. Sift flour, salt, soda and add All Bran. Fold into creamed mixtured. Add soaked bran. Fill greased tins ⅔ full. Cook at 400° 20-25 minutes. Store remaining batter in 1 gal. jar in refrigerator. It will keep 4-6 months. Do not stir after putting into jar. Spoon from top to fill tins.

CINNAMON BRAN MUFFINS

preparation time: 10 minutes
cooking time: 20 minutes
serves 8-10

Ann Glenn
Encore

1½ cup whole bran cereal
½ cup boiling water
1¼ cup flour
⅓ cup sugar
1¼ tsp. baking soda
¾ tsp. cinnamon

¼ tsp. salt
½ cup raisins
1 egg
1 cup buttermilk or lemon yogurt
¼ cup oil

Preheat oven to 425°. In large bowl, stir cereal and hot water; set aside to cool. In second bowl, stir together all dry ingredients including raisins. In third container, mix egg, yogurt and oil together. Combine liquid with cereal, then stir in dry ingredients. Mix thoroughly. (Store for future use; will keep 2 weeks or more in the refrigerator.) Bake about 20 minutes for regular sized muffins or about 10 minutes for mini-muffins — especially nice 'cause no one hesitates to "take two, they're small."

STICKY BUNS

preparation time: 10 minutes   *Gay Thompson*
cooking time: 20 minutes   *Satori*
serves 8

2 pkgs. Pillsbury Butterflake Rolls   **1 cup walnuts (chopped)**
Lots of butter or margarine   **1 cup raisins**
½ cup of honey

Melt 4 Tblsp. butter in large skillet. Separate rolls into individual pieces and fry on both sides (they'll puff up — each roll can be broken down into 3 to 4 pieces). Keep cooked pieces to one side so they don't burn and add butter as needed. When they're all cooked through, add raisins, honey and walnuts. Stir over heat until all are coated. This recipe was originally from Cooking on the Go, by Janet. I added the raisins.

BREAKFAST EASY SWEET ROLL

preparation time: 20 minutes   *Betsy Sacket*
cooking time: 30 minutes   *Insoucian*
serves 4-6

1 pkg. biscuits, unbaked (Ballard)   **½ cup walnuts, chopped**
1 can mandarin oranges   **¼ cup butter**
1 can pineapple chunks   **¼ cup brown sugar, packed**
                                         ½ tsp. cinnamon

Simmer butter, 2 Tblsp. each of mandarin and pineapple juices, brown sugar and cinnamon. Cut biscuits into quarters. Layer bread pan with ½ of biscuits, ½ of pineapple and mandarin oranges and ¼ cup walnuts. Repeat layering procedure. Pour slightly thickened butter sauce over biscuit mixture. Bake at 400° until brown about 30 minutes.

LUNCHEON

SHRIMP QUICHE

preparation time: 20 minutes
cooking time: 40 minutes
serves 8

Debbie Olsen
Kingsport

1 ready made fold-out pie crust
1 lb. frozen shrimp, cooked
1 can asparagus tips, drained
1 cup grated swiss cheese
2 cups cream

5 eggs
1 tsp. nutmeg
1 tsp. white pepper
½ tsp. salt
2 Tblsp. dried parsley

Line quiche pan with pastry. Arrange asparagus in wheel pattern in bottom of crust followed by a layer of shrimp. Combine remaining ingredients and pour on top. Bake for 40 minutes at 375° or until knife inserted in center comes out clean. *Serve with fruit salad.*

CHEESE SOUFFLE COCKAIGNE

preparation time: 15 minutes
cooking time: 25-30 minutes
serves 4

Adriane Biggs
Taza Grande

3 Tblsp. butter
3 Tblsp. flour
1 cup half-and-half cream
1 tsp. salt

¼ tsp. cayenne pepper
½ tsp. paprika
4 eggs (separated)
¼ lb. sharp cheddar cheese,
   grated

Preheat oven to 350°. Melt butter over low heat. Add flour and blend over low heat (3 to 5 minutes). Stir in slowly 1 cup cream. Allow to thicken. Then add grated cheese, beaten egg yolks, cayenne pepper, paprika and salt. Prepare one 7-inch souffle baker by buttering and dusting with flour. Beat egg whites until stiff, but not dry. Fold into cheese mixture and pour into souffle baker. Place in center of oven and bake for 25-30 minutes. Should be golden brown and puffed when done. *Serve with tossed salad and Vinaigrette Dressing.*

• *To loosen chilled crumb crust pie that sticks to the pan, wrap a hot, wet towel around the bottom and sides. Crust will loosen so each piece of pie will slip out easily.*

SALMON QUICHE

preparation time: 30 minutes
cooking time: 1 hr. 5 minutes
serves 6

Kandy Popkes
Wind Song Oregon

CRUST (or use 8″ frozen crust)
1 cup whole wheat flour
⅔ cup shredded sharp cheddar
    cheese
¼ cup chopped almonds
½ tsp. salt
¼ tsp. paprika
6 Tblsp. vegetable oil

FILLING
1 — 15 oz. can salmon
3 beaten eggs

½ cup crust mixture
1 cup dairy sour cream or cottage
    cheese
¼ cup mayonnaise
½ cup shredded sharp cheddar
    cheese
1 Tblsp. grated onion
¼ tsp. dried Dillweed
6-8 drops hot pepper sauce or
    Tabasco

FOR CRUST: Combine first 5 ingredients. Stir in oil and set aside ½ cup crust mixture. Press remaining mixture onto bottom and up sides of 9″ pie pan. Bake crust for 10 minutes in 400° oven. Remove from oven — reduce temperature to 325°.

FILLING: Drain salmon reserving liquid. Flake salmon removing bones and skin and set aside. In a bowl blend eggs, sour cream, mayo and reserved liquid. Stir in salmon, ½ cup cheese, onion, dillweed and hot pepper sauce. Spoon filling into crust. Sprinkle with reserved crust mix. Bake in 325° oven for 45 minutes or until firm in center.

REAL MAN QUICHE

preparation time: 15 minutes
cooking time: 45 minutes
serves 6

Jessica Adam
Nani Ola

10 slices bacon, crisp & crumbled
1 pie shell (baked at 450° for 10
    minutes)
4 eggs
2 cups light cream (or milk)
½ tsp. salt

⅛ tsp. nutmeg
Pinch of cayenne
1¼ cups Swiss cheese, grated
2 Tblsp. chopped onion

Preheat oven to 425°. Sprinkle bacon, cheese and onions over the shell. Combine remaining ingredients and beat thoroughly. Ladle over all. Bake 15 minutes at 425°; then lower to 350° for about 30 minutes more. *Serve hot or cold. Good with a salad for lunch or with fried potatoes and muffins for a brunch.*

CHEESE SAUSAGE BAKE

preparation time: 30 minutes   *Liz Thomas*
chilling time: 2 hours   *Raby Vaucluse*
cooking time: 1-1½ hrs.
serves 6

1 box seasoned croutons   2½ cups milk
1 — 16 oz. pkg. seasoned sausage 1 can cream of mushroom soup
2 cups grated cheddar cheese   ½ cup milk
4 eggs

Cover bottom of 9x13 oblong pan with croutons. Brown sausage and put on top of croutons. Sprinkle with cheese. Mix eggs and 2½ cups milk. Pour over all and refrigerate 2 hours or overnight. Mix soup and milk and pour over all. Bake 1 to 1½ hours at 300° (be sure it's done).

I usually prepare my own croutons by cutting up squares of bread in bottom of pan and baking with butter and garlic till done.

QUICHES, MEXI WAY

preparation time: 20 minutes   *Cheryl Anne Fowler*
cooking time: 30 minutes   *Vanity*
serves 4

4 6-inch flour tortillas   ½ cup wheat germ
4 ounces Monterey Jack cheese   2 cups skim milk
   with peppers, sliced   4 beaten eggs
1 medium onion, sliced and   ½ tsp. salt
   separated into rings   ½ tsp. chili powder
1 beaten egg   ¼ tsp. dry mustard

Gently press one flour tortilla in each of four individual au gratin casseroles; top with cheese slices. Dip the onion rings into the 1 beaten egg, then coat liberally with wheat germ. Top the cheese slices with about two-thirds of the onion rings; reserve remainder. In saucepan heat milk till almost boiling. Gradually add the milk to the remaining eggs, blending well; stir in salt, chili powder, and mustard. Place casseroles in shallow baking pan; place on oven rack. Divide egg mixture evenly among the casseroles. Top with reserved onion rings. Bake in a 350° oven about 30 minutes or till knife inserted off-center comes out clean. Let quiches stand at room temperature for 5 minutes before serving. Garnish quiches with parsley sprigs and pickled peppers, if desired.

SWISS ONION TART WITH CHEESE

preparation time: 40 minutes   *Louella Holston*
cooking time: 1½ hours   *Eudroma*
serves 8-10

Tart Pastry (recipe follows)
1 Tblsp. Dijon-style mustard
2 Tblsp. vegetable oil
5 cups thinly sliced yellow onions
   (about 1¼ lbs.)
½ tsp. sugar
¾ cup grated Swiss Gruyere
   cheese

¾ cup grated Swiss Emmenthal
   cheese
2 eggs
1 cup light cream or half-and-half
1 tsp. salt
⅛ tsp. freshly grated nutmeg

Make Tart Pastry

Roll pastry ⅛ inch thick on lightly floured surface. Fold into quarters; ease and unfold into 11 x 1-inch tart pan. Trim pastry flush with edge. Pierce bottom in several places with fork. Refrigerate 15 minutes.

Heat oven to 400°. Line pastry with aluminum foil; fill with dried beans or rice; place pan on baking sheet. Bake until pastry is set, about 12 minutes. Remove foil and beans. Brush mustard on bottom of pastry. Bake until pastry is cooked through and golden, 10 to 12 minutes. Cool on wire rack.

Heat oil in medium skillet over medium-high heat. Stir in onions and sugar. Cook, stirring occasionally until onions are soft and lightly golden, about 15 minutes. Cool to room temperature.

Mix onions and the cheeses; spread evenly in pastry shell. Whisk eggs lightly in small bowl; stir in cream and salt. Carefully pour over onion mixture. Sprinkle with nutmeg. Bake on baking sheet until custard is set and top is golden, 30 to 40 minutes. Cook on wire rack 10 to 15 minutes. *Serve warm.*
Tart Pastry

2 cups all-purpose flour
¼ tsp. salt

¾ cup lightly salted butter, cold
4-6 Tblsp. ice water

Mix flour and salt in medium bowl. Cut in butter until mixture resembles coarse crumbs. Gradually stir in water with fork until mixture cleans sides of bowl. Gather into ball; flatten slightly. Refrigerate, wrapped in plastic wrap, 30 minutes.

CRAB & SPINACH QUICHE

preparation time: 20 minutes   *Erica Benjamin*
cooking time: 40 minutes   *Caribe Monique*
serves 6

2 Extra large or 3 medium eggs   1 (10 oz.) pkg. frozen spinach,
1 can evaporated milk   cooked & drained
1 cup Muenster or Swiss cheese   1 — 9" pie crust
   (grated)   4 strips cooked bacon, crumbled
1 can (7 oz.) crab meat, drained

In pie crust shell, cover bottom with crab meat, spinach, and bacon. Add grated cheese. Pour egg and milk mixture on whole thing and cook. Bake at 375° approximately 40 minutes.

CHEESE PUDDING

preparation time: 15 minutes   *Saracen*
cooking time: 1 hour
serves 4

6 eggs, beaten   Mix together. Add:
½ cup milk   ½ cup flour
1 lb. jack cheese, cubed   1 tsp. baking powder
1 cup cottage cheese

Put in greased casserole. Bake 1 hour at 350°. Cut in squares. *Serve with salad.*

• *Make delicious, juicy hamburgers by adding one grated potato to each pound of ground meat.*

LINDA'S IMPOSSIBLE QUICHE

preparation time: 20 minutes
cooking time: 50-55 minutes
serves 4-6

Jan Robinson
Vanity

12 slices bacon (about ½ lb.)
   crisply fried and crumbled
1 cup shredded natural Swiss
   cheese
½ cup finely chopped onion

2 cups milk
⅓ cup Bisquick baking mix
4 eggs
Salt and pepper to taste

Heat oven to 350°. Lightly grease a 10-inch pie plate. Sprinkle bacon, cheese and onion evenly over bottom of pie plate. Place remaining ingredients in blender. Cover and blend on high speed for 1 minute. Pour into pie plate. Bake until golden brown and knife inserted in center comes out clean (50-55 minutes). Let stand five minutes before cutting.

PENNY'S TEMPTATION

preparation time: 20 minutes
cooking time: 10 minutes
serves 8

Pernilla Stahle
Adventure III

¼ lb. Gorgonzola cheese
1 large leek
1 cup heavy cream
1 pkg. (10 oz.) early peas, frozen

2 pkgs. egg noodles
1 lb. thinly-sliced ham
Spices — black pepper, paprika

Boil noodles as per directions on package. Fry sliced leek and ham strips. Add Gorgonzola, cream and peas. Simmer for 10 minutes. Pour over noodles in serving dish.

EASY STROGANOFF

preparation time: 15 minutes
cooking time: 2½ hours
serves 6

Maureen Stone
Cantamar IV

2½ lbs. round steak, cubed
2 (10 oz.) cans cream of
   mushroom soup
1 (10 or 16 oz.) can of mushrooms

1 envelope of Lipton Onion Soup
   Mix
1½ cups sour cream
2 Tblsp. Worcestershire Sauce

Combine all ingredients except sour cream in large casserole dish. Bake covered for 2½ hours at 350°. Stir occasionally. Stir in sour cream before serving. *Serve with buttered egg noodles and fresh green beans.*

MEXICAN LASAGNA

preparation time: 30 minutes   Samantha Lehman
cooking time: 30 minutes   Lady Sam
serves 6

2 lbs. ground beef
1 large onion
1 pkg. lasagna noodles, cooked
Sour cream, about 1 pint
1 — 1½ jars Ragu Spaghetti
    Sauce
1 chunk cherizo (to your taste) or
    chili powder

1 pkg. sharp cheddar cheese
1 pkg. Mozzarella cheese (or Jack
    cheese)
1 medium can whole or chopped
    chilies
1 large container cottage cheese
Parmesan

Fry ground beef, cherizo (or chili powder) and onion until slightly browned and drain off fat. In a greased tin, place one layer of meat, next a layer of chopped chilies (seeds to your hot taste), next a layer of 2 cheeses (sliced or shredded), next a layer of lasagna noodles, next a layer of cottage cheese and sour cream mixed together, next a layer of sauce. Repeat until all ingredients are used (at least 2 layers). Top with grated Parmesan cheese. Place in oven until cheeses melt, sauce bubbles and top slightly browned. Can be layered to your own taste. This is a flexible recipe! *Serve with heated corn tortillas smothered in butter and chilled Margueritas (to cool the palate).*

TORTELLINI VINAIGRETTE

preparation time: 35 minutes   Jean Thayer
chilling time: 2 hours   Paradise
serves 4-6

1 lb. frozen or packaged cheese-
    filled tortellini
2 cups broccoli flowerets
1¼ cup julienne carrots
1½ Tblsp. scallion

Dressing:
6 Tblsp. olive oil
2 Tblsp. + 1 tsp. white wine
    vinegar

1 tsp. dijon
⅜ tsp. basil
1½ tsp. parsley
¼ tsp. thyme
1 tsp. minced garlic
½ tsp. salt

Cook tortellini according to directions. Drain, rinse and reserve in large bowl. Steam broccoli and rinse. Steam carrots and rinse. Add vegetables and scallions to tortellini. Whisk together ingredients for dressing and pour over salad. Cover and refrigerate at least 2 hours. Remove from refrigerator 30 minutes before serving. To serve: garnish with halved cherry tomatoes and nicoise olives. *Serve with Parmesan Puffs.*

HAM CASSEROLE

preparation time: 20 minutes

cooking time: 45 minutes

serves 4

Jan Robinson

Vanity

6 oz. cooked rice
10 oz. broccoli, cooked
4 oz. mushrooms, sliced
1 cup grated cheddar cheese
1 lb. ham, chopped

1 can cream of celery soup
1 cup mayonnaise
2 Tblsp. mustard
1 tsp. curry powder
⅓ cup Parmesan cheese

Grease a 9 x 13″ pan. Mix together the soup, mayonnaise, mustard and curry powder. Layer rice, broccoli, ham, mushrooms and cheddar cheese. Pour soup mixture over. Sprinkle with the Parmesan cheese. Bake at 350° F. for 45 minutes.

PASTA PRIMAVERA

preparation time: 45 minutes

cooking time: 15 minutes

serves 6

Ci Ci

Bon Papa D

½ cup (1 stick) unsalted butter
1 medium onion, minced
1 large clove garlic, minced
1 lb. thin asparagus, tough ends
   trimmed, cut diagonally into
   ¼-inch slices, tips left intact
¼ lb. mushrooms, sliced
6 oz. cauliflower broken to florets
1 medium zucchini, sliced
1 small carrot halved and sliced
1 cup whipping cream

½ cup chicken stock
2 Tblsp. chopped fresh basil
1 cup tiny frozen peas, thawed
2 oz. prosciutto or cooked ham,
   chopped
5 green onions, chopped
1 lb. fettuccine or linguine, cooked
   al dente, thoroughly drained
1 cup freshly-grated Parmesan
   cheese

Heat wok or large deep skillet to medium-high heat. Add butter, onion, and garlic and saute until onion is softened, about 2 minutes. Mix in asparagus, mushrooms, cauliflower, zucchini and carrot and stir fry 2 minutes. (At this point remove several pieces of asparagus tips, mushrooms and zucchini for garnish). Increase heat to high. Add cream, stock and basil and allow mix to boil until liquid is slightly reduced, about 3 minutes. Stir in peas, ham and green onion and cook 1 minute more. Season to taste with salt and pepper. Add pasta and cheese, tossing until thoroughly combined and pasta is heated through. Turn into large serving platter and garnish with reserved vegetables.

• *A peeled potato in the refrigerator will absorb odors.*

LINGUINI WITH WHELK SAUCE

preparation time: 25 minutes
cooking time: 15 minutes
serves 4-6

Betsy Sackett
Insouciant

¼ cup butter
2 cloves garlic, minced
3 Tblsp. flour
2 Tblsp. parsley, minced
½ tsp. dried basil
1 can chicken broth

¼ cup onion, chopped
3 stalks celery, chopped
1½ to 2 cups whelks, cleaned
2 cups sour cream
Salt and pepper
1 pkg. (8 oz.) linguini pasta

In small saucepan saute garlic in butter until pale yellow. Add parsley, basil, flour, stir and set aside. In large saucepan simmer celery and onions in chicken broth until tender. Stir in whelks, sour cream and butter mixture. Stir well and add salt and pepper to taste. On medium-low heat (you should use an asbestos or simmering pad) heat sauce until very hot. Cool linguini as directed. Drain and spoon sauce over linguini.

CHICKEN, HAM, RICE SALAD

preparation time: 30 minutes
chilling time: 1-2 hours
serves 6-8

Candy Glover
Anemone

2 cups cooked chicken, boned and
    diced
1 cup cooked ham, diced
2½ — 3 cups cooked white rice
1 pkg. frozen English peas,
    cooked and cooled

1½ — 2 Tblsp. mayonnaise
Salt
Pepper

Mix all ingredients 1 to 2 hours before serving and place in refrigerator to cool. Place ½ cup of mixture on individual beds of lettuce. Garnish on the side with tomato wedges, black olives and white asparagus. *Serve with crackers or French bread.*

BLACK BEAN SPECIAL

preparation time: 10 minutes
cooking time: 20-30 minutes
serves 6-8

Jan Robinson
Vanity

3 cans black bean soup
1 cup onion, finely chopped
½ cup celery, finely chopped
3 large cloves garlic, finely
    chopped
½ cup olive oil
1 tsp. pepper

¼ tsp. ground cumin
1 Tblsp. oregano
Dash of Tabasco
3 cups cooked rice (saffron or
    white)
1½ cups onion, finely chopped

Saute onion, celery, garlic in olive oil. Add pepper, cumin, oregano and tabasco. Cook for a couple of minutes before adding the 3 cans of black bean soup. Mix thoroughly and simmer for 15 minutes or so. Serve soup in individual bowls with a large bowl of cooked saffron rice and another bowl of finely chopped onions for your guests to help themselves from. *I also serve a freshly baked loaf of bread and a fresh fruit and cheese platter. This makes a great quick lunch.*

BARBECUED WEINERS

preparation time: 15-20 minutes
cooking time: 45 minutes (or 15 minutes in Microwave)
serves 8-12

Ann Glenn
Encore

2 or 3 lbs. weiners, sliced
1 can carrot rings, drained
1 can tomato soup
4 red onions, chopped
2 large cloves garlic

1 green pepper, chopped
2 Tblsp. Worcestershire sauce
½ cup vinegar
½ cup barbeque sauce (or ketchup)
¼ cup vegetable oil

Combine all ingredients in large ovenproof bowl. Bake about 45 minutes at 350° or Micro about 15 minutes, or until thoroughly heated.

This receipe bears a close resemblance to the popular "Copper Pennies" and can also be enjoyed cold. I prefer to take it ashore hot in a 1-gal. Igloo thermos for beach parties. Also serve Zucchini Squares and Corn Confetti Salad.

TURKEY POT PIE

preparation time: 20 minutes
cooking time: 45 minutes
serves 8

Ann Glenn
Encore

5 Tblsp. butter
1 medium onion, diced
½ cup all-purpose flour
2½ cups milk
1 cup water
chicken boullion cube or pkt.
1 tsp. salt
¼ tsp. pepper

3 cups bite-sized chunks cooked
   turkey
3 red potatoes, thinly sliced
¼ lb. mushrooms, quartered
1 (10 oz.) pkg. frozen mixed
   vegetables
Pie crust mix

Make filling (stove top or microwave). Cook onion in butter until tender. Stir in flour until blended; cook 1 minute. Gradually stir in milk, water, salt, pepper and boullion. Cook, stirring constantly until mixture is slightly thickened. Stir in turkey, potatoes, mushrooms, and frozen mixed vegetables, breaking up vegetables with a fork. Spoon filling into 9x13" baking dish.

Prepare pie crust and roll out to 15x11" rectangle. With knife, cut out 6x3" rectangle to use for decorations (turkey, leaves, cornucopia, etc.). Cover turkey mixture and pinch edges. Decorate. Bake 45 minutes at 375°.

SHRIMP RAREBIT

preparation time: 20 minutes
cooking time: 10 minutes
serves 8

Adriane Biggs
Taza Grande

2 lbs. raw baby shrimp (shelled
   and deveined)
½ lb. cheddar cheese, grated
3 cloves garlic, crushed
1½ tsp. Worcestershire

4 tsp. butter
¼ tsp. cayenne pepper
1 tsp. salt
2 tsp. sherry
Toast points

Saute garlic in butter. Add remaining ingredients except shrimp. Cook, stirring constantly until cheese is melted and ingredients are blended. Add shrimp and cook about 2 minutes or until shrimp are done. Serve over toast points.

CORNISH PASTIE

preparation time: about 20 minutes   Adraine Biggs
cooking time: 45 minutes   Taza Grande
serves 4

Pastry:
8 oz. flour
4 oz. butter
4 Tblsp. cold water
Pinch salt
Filling:
1 lb. finely chopped beef or lamb

2 medium-sized finely chopped
   potatoes
1 finely chopped onion
1 cup chopped turnip (rutabaga)
Salt
Pepper
Basil

Roll out pastry on floured board to ¼ inch thick and cut into dinner plate size circles (4). On each circle place layer of potato, onion and turnip mix and place layer of meat on top. Season well. Fold over and crimp edges together well, make small slit in top to let out steam. Bake at 400° until pastry is golden and then reduce to 350° for about 40 minutes. Glaze with beaten egg or milk. Eat hot or cold. For extra taste try finely-sliced leeks! *Serve with a Spinach Salad and Hot Bacon Dressing.*

DANISH MEATBALLS

preparation time: 30 minutes   Jessica Adam
cooking time: 30 minutes   Nani Ola
serves 6

Meatballs:
1½ lbs. ground leaf
1 lb. ground pork
1 tsp. salt
⅛ tsp. pepper
1 egg
1 Tblsp. onion
½ cup milk
¾ cup bread crumbs
⅛ tsp. allspice

Dill Sauce:
¼ cup butter
¼ cup flour
2 cups chicken broth
¼ tsp. salt
2 tsp. dillweed
1 cup sour cream

Combine meatball mixture. Shape into balls. Broil for 20-30 minutes until browned. Prepare dill sauce; in saucepan melt butter, stir in flour. Stir in broth, salt and dillweed. Cook and stir over medium heat until thickened. Remove from heat; stir in sour cream and then pour over meatballs. *Can be served over egg noodles, accompanied by zucchini sauteed in garlic butter and Blueberry Muffins. Or, serve them alone as an appetizer.*

ONION PIE

preparation time: 15 minutes   Maureen Stone
cooking time: 15-20 minutes   Cantamar IV
serves 6

2 cups cottage cheese   1 baked 9-inch pastry shell
2 Tblsp. cream   1 (1 lb.) can stewed tomatoes,
⅛ tsp. salt   drained
⅛ tsp. pepper   ½ tsp. oregano
2 medium red onions, chopped   ½ tsp. salt

Mix cottage cheese with cream, salt, pepper, onions. Put cheese mixture
in pie shell. Top with drained tomatoes in a circle design; sprinkle with
oregano and ½ tsp. salt. Bake at 350° for 15-20 minutes. *Serve with
green salad.*

MOUSSAKA

preparation time: 20 minutes   Casey Wood
cooking time: 30 minutes   Megan Jaye
serves 6

17 oz. can corn, drained   2 slightly beaten eggs
1 lb. ground beef   1½ cup cottage cheese
1 Tblsp. flour   ¼ cup parmesan cheese
8 oz. can tomato sauce   4 oz. shredded Mozzarella
1 tsp. garlic salt   Slivered almonds
Dash of cinnamon and nutmeg

Spread corn in casserole dish. Brown meat, drain, add flour. Stir one minute. Add tomato sauce, garlic, cinnamon and nutmeg. Pour over corn.
Bake at 350° for 15 minutes. Combine eggs and cottage cheese, spread
over meat mixture. Top with Parmesan and Mozzarella. Sprinkle with
almonds. Bake 15 minutes more. Onions and green peppers can be added
to the meat while browning. *Serve with hot bread for soaking up the
sauce.*

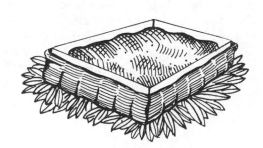

CHILI SIZE

preparation time: 10 minutes
cooking time: 10-15 minutes
serves 6

Samantha Lehman
Lady Sam

4 (8 oz.) cans Hormel chili (or
   other, if you like)
1½ lbs. ground beef
Onion

Cheese, very sharp
Hamburger buns, halved
Butter
Garlic, powdered or salt

In a pan heat your good store purchased chili. Make 6 hamburger patties out of the ground beef. Put patties on grill or under broiler. (Do not fry if possible, as this adds too much grease). Spread hamburger buns with butter and garlic and toast in your oven or grill. Grate about ½ cup of the cheese. Chop ¾ cup of onion. Layer: toasted bun, hamburger pattie, hot chili, sharp cheese and chopped onion.

I always prepare extra as I have found numerous requests for seconds (even though it is very filling and delicious). Slices of fresh pineapple are an excellent, and only, addition necessary.

HAM & BROCCOLI ROLLS

preparation time: 30 minutes
cooking time: 25 minutes
serves 8

Kandy Popkes
Wind Song Oregon

8 rectangular slices of boiled ham
   ¼" thick
8 slices Swiss cheese
3 (10 oz.) boxes frozen broccoli
   spears, cooked and drained or
   equivalent of fresh broccoli
2 Tblsp. butter or margarine
2 Tblsp. horseradish

2 Tblsp. flour
4 tsp. prepared mustard (Dijon)
½ tsp. salt
1 tsp. Worcestershire
1 tsp. grated onion
4 slightly beaten egg yolks
2 cups pineapple juice
1 cup milk

Top each ham slice with slice of cheese, then broccoli. Melt butter in heavy saucepan. Blend in flour, salt, horseradish, Worcestershire and onion. Combine egg and pineapple juice, blend into butter mixture, stir in milk and cook over low heat until thick and bubbly. Spoon about 1 Tblsp. sauce over broccoli. Roll ham and cheese around broccoli. Secure with toothpicks. Place rolls in shallow baking dish and cover. Bake in 350° oven about 25 minutes. Reheat sauce if necessary and spoon hot mustard sauce over ham rolls. Garnish with pineapple slices and spiced crab apples.

CONCH CHOWDER CARIBBEAN

preparation time: 30 minutes   *Paula Taylor*
cooking time: 2 hours   *Viking Maiden*
serves 12

½ lb. bacon, chopped
1 head celery, chopped
3 large onions, chopped
5 lbs. conch, tenderized twice and
   chopped
1 pint double cream
½ gal. milk

2 cubes chicken stock
3 Tblsp. flour to thicken
Salt & pepper
Dash HOT West Indian pepper
   sauce
1½ lbs. potatoes, diced

Saute bacon in large deep pan, add chopped celery, onions and conch. Turn over a few times to lightly brown ingredients. Add milk, stock cubes, potatoes, salt and pepper, bring to boil. Lower heat and simmer for 2 hours or more. This part can be done in pressure cooker, bring to steam and simmer for ½ hour, cool and remove lid, proceed to next step — add flour to double cream and gradually add to chowder. This makes the chowder thick and creamy. Just before serving add dash of West Indian Hot Sauce. *Serve with heated French bread or West Indian Coconut Bread and butter. The West Indian Sauce should be served as a condiment for those who like really hot, spicy food.*

FISH CHOWDER

preparation time: 30 minutes   *K. A. Strassel*
cooking time: 30 minutes   *Great Escape*
serves 6

2 lbs. fish chunks (use strong fish
   — kingfish, bonita, etc.)
1 cup diced bacon
3 cups diced potatoes
1 cup sliced carrots
1 cup sliced celery

1 chopped onion
1 clove garlic — pressed
2 cups milk
salt, pepper, Worcestershire, basil,
   thyme, oregano

Peel and dice potatoes, steam until tender, drain. Saute bacon until almost crisp, remove, saving drippings. Saute vegetables in bacon drippings until tender (add water if necessary), remove. Saute fish in butter until white, remove. Combine fish, potatoes, vegetables, bacon, milk and seasonings to taste. Simmer as long as you like (at least ½ hour) until flavors blend. *Serve with large green salad and garlic bread.*

WATER SAFARI SPECIAL

preparation time: 15 minutes   *Marty Peet*
cooking time: 45 minutes   *Grumpy III*
serves 6

1 onion, chopped
3 stalks celery, chopped
2 cloves garlic, minced
2 Tblsp. butter
½ cup parsley, chopped
2 lb. can tomatoes
2-3 cups V-8 juice or tomato
½ cup white wine

1 cup water
1 green pepper, chopped
3 bay leaves
1 tsp. each of basil and oregano
2 tsp. horseradish
1 tsp. Dijon mustard
salt and pepper
1½ lbs. dolphin or any chosen fish

In a Dutch oven, saute onion, celery and garlic in butter. Add all the other ingredients, except the fish and simmer with lid on at least ½ hour. Cut fish into 1-inch chunks, add and simmer 15 minutes more. *Serve with a huge salad and crusty French or garlic bread.*

PETER ISLAND BROCCOLI HAM CASSEROLE

preparation time: 25 minutes   *Frances Bryson*
cooking time: 20 minutes   *Amoeba*
serves 6

1 pkg. frozen broccoli
1 cup chopped, cooked ham
1 Tblsp. parsley
2 Tblsp. green pepper, chopped
2 hard-cooked eggs, chopped
¼ cup grated American cheese
1 tsp. onions, finely chopped

4 tsp. lemon juice

Cream sauce:
1½ Tblsp. butter
1½ Tblsp. flour
½ Tblsp. salt
1½ cups milk

Mix ham, parsley, green pepper, eggs, cheese, onions, and lemon juice. Cook broccoli and cut into 1-inch pieces, do not overcook. Place broccoli in buttered casserole, cover with ham mixture. Cover with cream sauce, sprinkle with ½ cup buttered bread crumbs. Bake for 20 minutes at 350°.

Cream Sauce: Melt butter and mix in flour. Slowly blend milk. Stirring constantly, heat until sauce becomes thick and creamy.

• *A little salt sprinkled in the frying pan will prevent fat from spattering.*

PIZZA

preparation time: 30 minutes
rising time: 1½ — 2 hours
cooking time: 20 minutes
serves 6

Kyle Perkins
Saracen

½ cup warm water
1 pkg. yeast
1 tsp. sugar
2 cups flour

⅓ cup water
2 tsp. olive oil
1 tsp. salt

Mix first 3 ingredients and let sit for 2 minutes. Stir, and let sit 5 more minutes. Add flour and remaining ingredients. Mix well, knead, let rise 1½ — 2 hours.

Sauce:
8 oz. tomato sauce
6 oz. tomato paste
¼ tsp. oregano

¼ tsp. thyme
¼ tsp. Rosemary
¾ tsp. garlic salt
¼ cup water

Mix, simmer, let cool. Spread dough on greased and floured cookie sheet. Add sauce, meat, mushrooms, etc. Then cheeses.

Grated Cheeses — ¼ lb. Mozzarella and ¼ lb. Monteray Jack

Bake at 425° for 20 minutes.

SATORI PIZZA

When the moon hits your eye like a big pizza pie, that's Satori!
preparation time: 20 minutes
cooking time: 10-15 minutes
serves 8

Gay Thompson
Satori

1¼ cups Bisquick
1 egg
¾ cup mayonnaise
4 oz. pepperoni skinned and sliced
2 cups shredded mozzarella
½ cup sliced black olives

½ cup chopped green pepper
¼ cup chopped green onion
4 Tblsp. capers
½ tsp. Italian seasoning
¼ tsp. garlic powder
8 slices rye bread

Mix mayonnaise, egg and stir in bisquick. Add other ingredients, mix well and spread onto the bread. Bake on an ungreased cookie sheet for 10 minutes or until puffy golden brown. *Serve with a tossed green salad and an Italian red wine.*

CARIBBEAN BOULIABAISSE

preparation time: 10 minutes   *Kathy Wagner*
cooking time: 40 minutes   *Planktos*
serves 4

½ lb. Linguica or Italian sausage
⅛ cup olive oil
½ cup chopped onion
1 green pepper, chopped
2 garlic cloves, chopped
¼-½ lb. prosciutto or ham,
 chopped
1 can peeled tomatoes
1 bay leaf

Salt and pepper
1½ cups water
¼ cup wine
2 tsp. lemon
Steamers, mussels or small
 cherrystones
White fish cut in 2″ pieces
Shrimp or Lobster

Remove sausage from casing, drop in sieve and plunge in boiling water for 1 minute. Saute onions, garlic and pepper in olive oil approximately 5 minutes. Add sausage, ham, tomatoes, bay leaf, salt and pepper and cook briskly until liquid boils off and mixture holds a spoon shape. Add water, wine and lemon and bring to boil stirring regularly. Reduce heat and add steamers, mussels, or small cherrystones and cook covered 10 minutes. Add fish, shrimp, and lobster and cook 5 minutes. *Serve with crusty garlic bread.*

SEAFOOD MOUSSE

preparation time: 20 minutes   *Debbie Olsen*
chilling time: 3 hours   *Kingsport*
serves 8

2 envelopes unflavored gelatin
1 (15 oz.) can red salmon
1 — 1 lb. snow crab meat (thawed
 and drained)
2 cups mayonnaise
½ cup chili sauce
2 Tblsp. lemon juice

¼ cup capers
1 Tblsp. Worcestershire sauce
1 tsp. dillweed
½ tsp. white pepper
3 hard-cooked eggs, chopped
½ cup green olives, sliced

Bone and drain salmon, reserving liquid. Add water if necessary to make ½ cup. Combine with gelatin and heat to dissolve. Set aside. In a mixing bowl, gradually blend gelatin with mayonnaise. Combine ½ of crab meat and the remaining ingredients, except capers. Pour into a ring mold and chill at least 3 hours. After unmolding, put remaining crab meat in center and sprinkle capers around the top of mousse. Beautiful and delicious. *Serve with freshly baked Beer Bread and a green salad.*

JELLIED SALMON MOUSSE

preparation time: 40 minutes   *Margaret Benjamin*
chilling time: 2-3 hours   *Illusion II*
serves 4-6

1 packet gelatin	8 oz. carton cottage cheese
¼ pt. cold water	1 green pepper
Juice of ½ lemon	2 Tblsp. chives or parsley
1 lb. can red salmon	1 tsp. salt and cayenne pepper
¼ pt. mayonnaise	½ cucumber sliced for garnish

Sprinkle gelatin on cold water in pan and leave until dissolved. Warm over low heat until clear. Do not boil. Add lemon juice and cool. Mix salmon, cottage cheese and mayonnaise. Add green pepper, parsley and seasonings. Pour 2-3 Tblsp. jelly into bottom of mold or bread pan. Place a few slices of cucumber in jelly and put into refrigerator to set. Blend remaining jelly into salmon mixture and pour into mold. Leave to set. *Serve with a green salad and hot garlic bread. Followed by Coffee Souffle.*

LOBSTER MOUSSE

preparation time: 20 minutes   *Pamela McMichael*
chilling time: 2-3 hours   *Pieces of Eight*
serves 8-10

2 pkgs. lemon jello	3 Tblsp. chopped pimentos
1 cup cold water	½ cup green pepper strips
1 cup Miracle Whip	1 tsp. grated onions
4 Tblsp. lemon juice	2 cups hot water
2 cups lobster	½ tsp. salt
½ cup sliced stuffed olives	1½ cups diced celery

This salad looks particularly attractive when served in a fish shape mold. The contours of the fish lend themselves beautifully to decoration.

Dissolve jello in hot water, add cold water, lemon juice, Miracle Whip, salt and onion. Blend with rotary beater. Pour into bowl and refrigerate until partially jellied (about ½ hour). Oil salad mold. Decorate bottom of mold while waiting, using pimento strips, green pepper strips, olives. Remove gelatin mixture from refrigerator and whip till fluffy. Fold in lobster and celery. Pour into decorated mold and chill.

TURKEY CITRUS SALAD

preparation time: 15 minutes (plus chilling)   *Ann Glenn*
cooking time: about 5 minutes   *Encore*
serves 6-8

3 Tblsp. flour
3 Tblsp. sugar
½ tsp. salt
1 egg
1 cup milk
½ cup orange juice
4 cups cubed cooked turkey (or
    chicken)

2 cups celery, thin sliced
1 cup toasted slivered almonds
1 cup (about 3 medium) fresh
    orange sections, or 1 can
    Mandarin sections

Make cooked dressing first. Combine flour, sugar and salt in medium saucepan. Blend in egg. Gradually stir in milk and orange juice. Cook over medium heat stirring constantly until mixture just comes to a boil and thickens. Chill, stirring occasionally.

Combine turkey, celery, almonds, orange sections and dressing. Mix lightly but thoroughly. Chill before serving on lettuce leaves. *Serve with Hasty Chowder and fruit with Amaretto Cream.*

CARIBBEAN CHICKEN SALAD

preparation time: 40 minutes   *Lori Moreau*
cooking time: 15 minutes   *Alberta Rose*
serves 6

1 cup dry rice
2 cups cooked chicken, shredded
¾ cup raisins
¾ cup mango chutney

1 large pineapple
Mayonnaise
Cinnamon

Cook the rice, then cool. Cut the pineapple in half, including flower. Scoop out meat and cut it into chunks. Save shells. Combine pineapple chunks and remaining ingredients and rice. Mound into pineapple shells and garnish flower with orange twists and marachino cherries. *Excellent served with Chilled Cucumber Salad and fresh Beer Bread.*

SMOKED CHICKEN APPLE WALNUT SALAD

preparation time: 20 minutes   *Kathleen McNally*
chilling time: about an hour   *Orka*
serves 6

4 whole smoked chicken breasts
(about 2 lbs.), skinned, cut
into thin strips
3 medium Granny Smith's apples
(cored and diced)
6 celery stalks, sliced (1½ cups)
4 cups chopped watercress leaves

1 cup lemon-mustard dressing
(recipe follows)
Freshly ground pepper
Boston or Romaine lettuce leaves
Watercress sprigs (garnish)
1¼ cups chopped toasted walnuts
(5 ounces)

Toss chicken strips, apple, celery and chopped watercress in large bowl. Blend in Lemon Dressing and pepper. Cover and chill before serving.

To serve, line platter with lettuce leaves. Mound salad in center. Garnish with watercress sprigs. Sprinkle with nuts.

(Smoked Turkey Breast can be substituted)

LEMON-MUSTARD DRESSING — Makes 1 Cup

4 tsp. fresh lemon juice
4 tsp. Dijon mustard
1 egg yolk
¼ tsp. each salt and freshly
ground pepper

1 cup olive oil
1 Tblsp. fresh lemon juice

Combine 4 tsp. lemon juice with mustard, egg yolk, salt and pepper. Blend in olive oil with rotary beater until thickened and mixed well. Blend remaining lemon juice. Add to salad ingredients, toss gently. *Serve White Wine Onion Soup as a starter with Gruyere Garlic Toast. Follow the salad with Pear Cranberry Crisp.*

HAM-STUFFED AVOCADOS

preparation time: 20 minutes   *Lori Moreau*
serves 6   *Alberta Rose*

1½ cups crushed crackers
2 hard-boiled eggs, chopped
½ cup chopped onion
¼ tsp. pepper

1½ – 2 cups ham cubed
mayonnaise
3 large avocados

Combine crackers, eggs, onion, pepper and ham. Add mayonnaise. Half and peel avocados and stuff with mixture. *Serve on lettuce beds.*

COLD CURRIED SALAD

preparation time: 45 minutes   *Barbara Tyne*
cooking time: 20 minutes   *Anodyne*
serves 4-6

1 lb. small shrimp, cooked and   ½ cup peanuts or cashews
   shelled   2 tsp. salt
1 apple, cored and chopped   ⅛ tsp. pepper
½ green pepper, seeded and cubed   ⅔ cup mayonnaise
4 scallions, chopped   ⅓ cup sour cream
1 ripe avocado, peeled, seeded   1 Tblsp. curry powder
   and cubed   ¾ cup cooked rice

In a large bowl combine shrimp, apple, green pepper, scallions, avocado, peanuts or cashews, salt and pepper. Mix together mayonnaise, sour cream and curry powder. Gently toss one-half the dressing with shrimp mixture. *Serve salad on a bed of cold rice with the remaining sauce on the side. Serve with Black Bread.*

This is also delicious using cooked, cubed chicken instead of the shrimp.

SZECHUAN SHRIMP SALAD

preparation time: 30 minutes   *Shirley Fortune*
chilling time: 1-2 hours   *Maranatha*
serves 6

1 pkg. fine cooked egg noodles   1 pkg. frozen snow peas, cooked 1
¼ cup vegetable oil   minute
2 Tblsp. vinegar   1 red bell pepper, cut in match
2 Tblsp. soy sauce   stick strips
1 Tblsp. sesame seeds   ½ cup scallions, sliced including
1 tsp. ground ginger   green ends
1 tsp. sugar   3 bananas cut diagonally in ½-
1 tsp. tabasco or to taste   inch slices
1 lb. small frozen, cooked shrimp

In large tupperware bowl combine oil, vinegar, soy, seeds, ginger, sugar, red pepper and hot sauce. Add shrimp, snow peas and scallions. Toss to coat. Chill covered until serving time. To serve: place cooked noodles on platter and gently mix sliced bananas into shrimp mix and spoon over noodles. This dish looks stunning and has a spicy oriental flavor. *Serve with hot croissants and chilled Soave Bolla wine.*

CONCH SALAD

preparation time: 30 minutes
marinating time: 3-4 hours
serves 6-8

Jan Robinson
Vanity

8 fresh conch, tenderized twice	Pepper
1 onion, chopped	Juice of 6 fresh limes
2 cloves garlic, minced	Tabasco to taste
2 green pepper, chopped	2 Tomatoes, diced
1 Tblsp. pimento, chopped	2 Limes, garnish

Dice conch. Place in bowl with onion, garlic, green pepper and lime juice. Marinate in the refrigerator for 3-4 hours. Just before serving, add tomatoes, pimentos and tabasco. *Serve on a bed of lettuce with a wedge of lime along with black pumpernickel bread or crackers.*

TUNA TOMATO

preparation time: 30 Minutes (plus chilling)
serves 6

Jean Thayer
Paradise

6 Medium tomatoes	1 Tblsp. raisins
3 (6½ oz.) cans tuna, drained	Mayonnaise
2 Anchovy fillets — mashed	Capers
1 Tblsp. dijon style mustard	Lemon wedges

Slice off tops of tomatoes, scoop out and drain. In mixing bowl, blend tuna, mashed anchovies, mustard, raisins and enough mayonnaise to satisfy your taste. Add salt and fresh ground pepper. Chill.

To serve, fill tomatoes, mounding high. Garnish with drained capers and a squeeze of a lemon wedge. *Serve with a hot soup and rolls, or Gougère.*

AVOCADO DELIGHT

preparation time: 20 minutes
serves 4

Kandy Popkes
Wind Song Oregon

4 avocado halves	1 bottle Thousand Island Dressing
1 small can grapefruit pieces	¼ lb. small shrimp (cooked)
1 pkg. crab stuffing	

Lightly cover avocado halves with lemon juice to prevent browning.

Place each half on plate with lettuce leaf. Fill hollow with grapefruit pieces. Mound crab stuffing on top. Pour Thousand Island dressing over mound. Place shrimp on top in design. Garnish plate with fresh fruit.

CHICKEN CREPES

preparation time: 30 minutes   Candy Glover
cooking time: 30 minutes   Anemone
serves 8

2 cups cooked chicken breast,   ¼ – ½ cup evaporated milk
    diced   Dash of sweet sherry
½ cup fresh mushrooms, diced   ½ Tblsp. tarragon
¼ cup green onion, diced   16 cooked crepes
1 chicken bouillon cube
¼ cup water

Melt chicken bouillon cube in water in skillet. Mix in all other ingredients
and saute until it thickens, stirring occasionally (about 15-20 minutes). Set
aside. (Now make or reheat Crepes). Fill each crepe with 1 large Tblsp. of
chicken filling. *Serve with a light chicken sauce or white sauce. Garnish
with parsley.*

GOURMET TOAST

preparation time: 15 minutes   Pamela McMichael
cooking time: 4-6 minutes   Pieces of Eight
serves 1

1 slice French bread   1 slice Swiss cheese
4 very thin slices dill pickle   1 oz. dry white wine
4 very thin slices tomato   Pepper
1 slice ham   2 Tblsp. butter

Pan fry both sides of bread in the butter to a golden brown. Place in baking dish or pan. Cover first with pickles, then tomatoes, ham and finally
cheese. Pour wine over the whole thing and bake or broil until cheese
melts evenly. Grind black pepper over top to taste.

POCKET HEROES

preparation time: 20 minutes   Different Drummer
marinating time: 2-3 hours

Pita Bread   Vegetables (onion, celery, tomato,
Leftover meat (bologna, ham,   pepper, etc.)
    salami, etc.)   Lettuce
Cheese   Oregano

Chop all of the above. Marinate from after breakfast until lunch time in
Italian dressing. Stuff into pita halves. Sprinkle with oregano. Enjoy.

CREPES VANDA

preparation time: 45 minutes   *Jeannie Drinkwine*
cooking time: 20 minutes   *Vanda*
serves 8

3 cups crab meat (about ¾ lb.)
¼ cup thinly sliced green onions
¼ cup butter
¼ cup flour
2 cups light cream
½ tsp. salt
2 tsp. soy sauce

2 Tblsp. dry white wine
2 tsp. finely chopped canned
  green chilies
2 avocados, thinly sliced
16 warm, cooked crepes (your
  favorite recipe)

Saute onions lightly in butter. Blend in flour and cook until bubbly. Stir in cream and salt. Cook, stirring constantly until thickened. Blend in soy sauce, wine, green chilies, and crab meat. Fill warm crepes with crab mixture; fold over. Garnish with avocado slices. *Serve immediately. Makes 16 crepes.*

HOT CRAB BOATS

preparation time: 30-40 minutes   *Lori Moreau*
cooking time: 15 minutes   *Alberta Rose*
serves 6

1 lb. lump crab meat
1 cup diced celery
1 cup cooked peas
6 oz. Swiss cheese grated

¼ cup chopped parsley
Mayonnaise
¼ cup melted butter
6 long hero rolls or 10 small rolls

Combine crab, celery, peas, cheese, parsley and mayo. Mix well. Slice tops from rolls and scoop out center to form a shell. Save tops. Brush inside of each shell with butter and fill with crab mixture. Replace tops. Brush outside of roll with butter and wrap individually in foil. (Crab boats may be refrigerated at this point and baked later.) Bake at 400° for 15 minutes. *Serve with a fruit salad and chips.*

CHICKEN AND CHUTNEY PITA BREADS

preparation time: 20 minutes   *Marty Peet*
cooking time: 15-20 minutes   *Grumpy III*
serves 4-6

6 Pita breads   **½ cup sour cream**
2 cups cooked chicken, diced   **½ cup chutney**
½ cup diced celery   **½ cup raisins**
1 cup diced Monterey Jack cheese   **1 tsp. curry powder**
3 Tblsp. mayonnaise

Preheat oven to 350°. Cut each Pita bread in half to fill. If Pita needs opening, put halves in tin foil and sprinkle with water. Pinch foil closed and warm in oven 5 minutes. They will puff open ready for stuffing. Mix all other ingredients together and spoon into Pita halves. Set in a baking dish, open side up. Cover with foil and heat 15 minutes. Arrange to form a circle around a bowl of Fruity Rice Salad.

GRILLED TOFU SANDWICHES

preparation time: 10 minutes   *Ginger Outlaw-Fleming*
cooking time: 3-5 minutes   *Coyaba*
serves 2

Pita bread   **Tofu**
Tomato slices   **Mayonnaise**
Alfalfa sprouts or lettuce   **Sesame oil**
Avocado   **Tamari soy sauce**

Slice tofu into sandwich size pieces. Fry in equal amounts of oil and tamari, about 1 Tblsp. each. Turn when lightly browned on one side and cook other side about 3-5 minutes on medium heat. Put in sandwich along with slices of salad ingredients. It is a nice low calorie alternative to a cheese sandwich and better for you, too.

TOFU is a soybean product and is cholesterol free, high in protein and very low in calories. It can be used in a lot of ways, or take the place of high-calorie food like cheese. I prefer it cooked with soy sauce. In St. Thomas you may buy it at "Vege-table" or at the "Fruit Bowl".

GRUMPY MELTS

preparation time: 15 minutes   *Marty Peet*
cooking time: 5 minutes   *Grumpy III*
serves 4

2 (7 oz.) cans chicken or tuna   2 Tblsp. Parmesan cheese
2 Tblsp. mayonnaise   Salt and pepper
¼ cup chopped celery   Paprika
¼ cup chopped chives or onion   8 slices Swiss cheese
Dash dill   4 English muffins, split

Preheat oven to 450°. Mix all ingredients, except cheese and muffins.
Spread mixture on muffin halves, top with cheese and bake or broil until
bubbly. Sprinkle on a dash more paprika and dill. Arrange on a serving
platter. *Serve with a rice or large green salad.*

SATORI BATH'S SANDWICH

preparation time: 15 minutes   *Gay Thompson*
serves 8   *Satori*

1 lb. canned ham   1 (16 oz.) can pitted black olives
6-8 oz. cream cheese, softened   16 slices pumpernickel bread
1-2 Tblsp. Dijon mustard   1 container alfalfa sprouts

Spread cream cheese on 8 of the bread slices, mustard on the other 8.
Slice ham ⅛″ thick and place on the cream cheese. Top with sprouts,
squeeze together with the ham slice. Pack in a picnic basket and take
with you to the Baths in Virgin Gorda. Add some potato chips and your
favorite beverage. They are very filling.

HOT CRAB SANDWICH

preparation time: 15 minutes   *Gay Thompson*
cooking time: 20 minutes   *Satori*
serves 8

8 English Muffins, split   1 Tblsp. horseradish
2 large tomatoes   4 dashes Tabasco
3 (4½ oz.) can Alaskan crab or 12   ½ tsp. Worcestershire sauce
    oz. King crab frozen   4-8 oz. grated cheddar cheese
¾ pint sour cream

Slice each tomato into 8 (⅛″) slices. Mix crab, sour cream, horseradish,
tabasco and worcertershire sauce together. (Add more tabasco if desired).
Toast muffins, top with tomato and pile with crab mixture. Sprinkle with
cheddar and bake or broil until the cheese melts and mixture is warmed
thru. These are very filling.   Great with a bottle of chilled white wine.

CROQUE MONSIEUR
(Dressed Up Ham & Cheese)

preparation time: 20 minutes
cooking time: 20 minutes
serves 6

Jean Thayer
Paradise

1½ cup grated Swiss cheese
10 Tblsp. heavy cream
Salt and pepper to taste
12 Slices firm white bread

6 slices ham — thin, trimmed to
   fit bread
3 eggs
Butter

Mix grated cheese with 6 Tblsp. heavy cream to make a spread. Season with salt and fresh ground pepper. Spread cheese mixture on one side of all bread slices. Top 6 slices with ham and cover with remaining cheese covered slices, to make a sandwich. Beat together eggs and 4 Tblsp. cream, season with salt and pepper. Dip each sandwich into egg mix. Melt 2 Tblsp. butter in fry pan over moderate heat, being careful not to allow butter to burn. Fry sandwich turning once. Hold in 200° oven 'til all have been cooked. Slice and serve.

POLYNESIAN CHICKEN SANDWICH

preparation time: 30 minutes
cooking time: 5 minutes
serves 4

Saracen

1½ lbs. Chicken Breasts, cooked
   and diced
½ cup mayonnaise
1 tsp. Dijon mustard
¼ cup finely sliced scallions
½ cup finely sliced celery
¼ tsp. salt

¼ tsp. curry powder
1 pkg. sliced almonds
1 (8 oz.) can sliced pineapple (or
   fresh is best)
½ cup shredded cheddar cheese
4 English Muffins or sourdough
   buns

Mix first 7 items and chill until ready to serve. Split and toast muffins, lightly butter. Place one slice of pineapple on each muffin half. Stir almonds into chicken mixture and spoon over pineapple. Sprinkle cheese over and broil or micro until hot and the cheese melts.

PARMESAN PUFF

preparation time: 30 minutes
cooking time: 45-55 minutes
serves 6

Jean Thayer
Paradise

⅔ cup water
⅓ cup milk
4 Tblsp. unsalted butter
¾ tsp. salt
¼ tsp. nutmeg

Dash of pepper
1 cup all-purpose flour
4 eggs
1 cup parmesan, grated

Preheat oven to 375°. In large saucepan combine water, milk, butter, salt, nutmeg and pepper. Bring to boil over medium-high heat. Remove from heat, add all of the flour and beat vigorously for 1 minute. Return to heat and continue to beat flour mixture vigorously for 2 minutes. (Great for upper arm muscles.) Remove from heat, cool slightly and add eggs one at a time, mixing till smooth and thoroughly blended. Stir in ½ of cheese mix to batter. Grease a cooking sheet and with large spoon make 7 equal mounds arranged in a circle, touching. Top with remaining cheese. Bake in the middle of oven 45-55 minutes, till puffed and golden. *Serve immediately with butter. (Goes with Tortellini Vinaigrette or any luncheon salad.)*

SHRIMP EGG ROLLS

preparation time: 30 minutes
cooking time: 30 minutes
serves 6

Chris Kling
Sunrise

1 pkg. egg roll wrappers
2 cans Chinese vegetables, drained
2 cans medium shrimp
½ cup chopped celery

3 Tblsp. soy sauce or more to taste
2 Tblsp. cornstarch
Oil for deep frying
Seafood cocktail sauce

Mix vegetables, celery and shrimp in a bowl. Combine soy sauce and cornstarch and pour over vegetables and shrimp. Stir until all are covered with sauce. Place about four tablespoons of the shrimp vegetable mixture on each egg roll wrapper. Fold up the sides and then roll. Seal closed with a mixture of cornstarch and water. Deep fry until golden brown and serve with seafood cocktail sauce for dipping. *Serve for lunch with fruit salad and fortune cookies.*

SALADS AND DRESSINGS

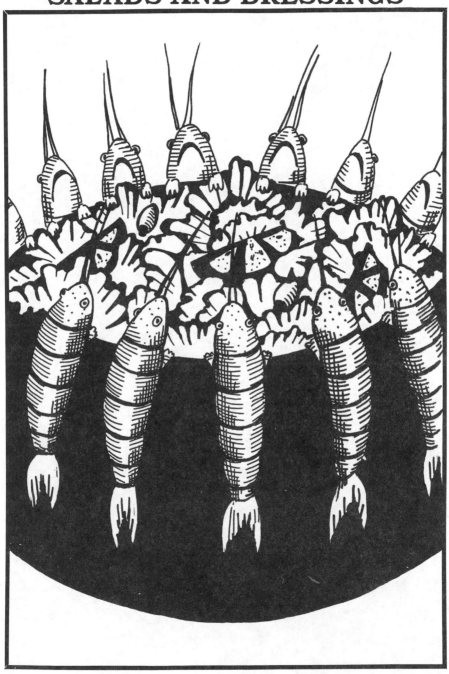

KOREAN SALAD

preparation time: 15 minutes
chilling time: 1-2 hours
serves: 8

Kandy Popkes
Wind Song Oregon

1 lb. fresh spinach
1 #2 can bean sprouts
8 slices of bacon, fried crisp and
   crumbled
1 can mandarin oranges
1 cup salad oil
¾ cup of sugar

⅓ cup ketchup
¼ cup vinegar
1 Tblsp. Worcestershire sauce
1 medium onion, finely chopped
Salt, to taste

Toss cleaned spinach pieces with bean sprouts, bacon and mandarin oranges. Mix all other ingredients to make dressing in advance and toss just before serving.

FRESH SPINACH SALAD

preparation time: 10 minutes (plus chilling)
cooking time: 5 minutes
serves: 4

D. J. Parker
Trekker

6 Tblsp. vegetable oil
2 Tblsp. wine vinegar or fresh
   lemon juice
¼ tsp. dry mustard
1 garlic clove, minced and
   crushed
Salt and pepper, to taste

6 cups fresh spinach, stems
   removed, torn in bite-size
   pieces
2 large tomatoes, cut in wedges
6 slices bacon, fried crisp

Mix first five ingredients in jar with lid and chill. Shake well before using. Toss with spinach and tomatoes. Arrange in bowls and sprinkle with crumbled bacon.

• *Buy the whitest endive as traces of green at tips indicates the endive has been exposed to the light and will be excessively bitter.*

SPECIAL SPINACH SALAD

preparation time: 12 minutes
cooking time: 5-10 minutes
serves: 8

Michelle Sutic
Harp

1 Tblsp. shallots, chopped
3 Tblsp. red wine vinegar
4 Tblsp. Worcestershire sauce
3 strips bacon, chopped
3 Tblsp. honey
1 lemon

1 lb. fresh spinach, cleaned
6 medium size mushrooms, sliced
Garnish:
Cherry tomatoes
Sliced hard boiled eggs

Cook bacon until half done. Sprinkle with shallots and cook about 1 minute. Add 4 dashes of Worcestershire and let cook for 30 seconds. Add 3 Tblsp. of vinegar and simmer 30 seconds. Then add the honey. Simmer everything for 60 seconds. While simmering, toss spinach and mushrooms and squeeze lemon over them. Pour dressing over salad and toss. Garnish with cherry tomatoes and hard-boiled egg. *Serve immediately.*

SPINACH SALAD

preparation time: 20 minutes
chilling time: 1 hour
serves: 6

Kathi Strassel
Great Escape

1 pkg. fresh spinach
6 slices cooked bacon, crumbled
½ cup sliced mushrooms
3 Tblsp. scallions, chopped

3 hard-boiled eggs, chopped
4 oz. Swiss cheese, slivers
Creamy vinaigrette dressing
Croutons (optional)

Wash and drain spinach, tear into manageable pieces, discarding stems. Add all other ingredients except dressing and croutons. Can chill for an hour. *Before serving, add dressing and toss.*

Dressing:
1 coddled egg
2 Tblsp. salad vinegar
6 Tblsp. safflower oil
Salt

Pepper
1 Tblsp. parmesan cheese
⅛ tsp. dry mustard
⅛ tsp. basil

In shaker, mix all the ingredients.

WILTED SPINACH SALAD

preparation time: 15 minutes   *Harriet Beberman*
cooking time: 5 minutes   *Avatar*
serves: 4-6

1 lb. fresh spinach
½ cup green onions
Dash ground pepper
5 slices bacon, diced
1 clove of fresh garlic
2 Tblsp. wine vinegar
1 Tblsp. lemon juice
1 tsp. sugar

½ tsp. salt
1 hard cooked egg, coarsely
    chopped
Garnish:
Mushrooms
Tomatoes
Egg quarters

Wash spinach and discard stems. Pat dry on paper towel, tear into bowl. Add onion and sprinkle with pepper. Chill. At serving time, slowly fry bacon and garlic in deep chafing dish, skillet, or wok till crisp. Add vinegar, lemon juice, sugar and salt. Gradually add spinach, tossing just till leaves are coated and wilted slightly. Sprinkle with egg. Garnish with mushrooms, egg quarters, and sliced tomatoes.

SPINACH SALAD CHEZ CARRE

preparation time: 30 minutes   *Louella Holston*
cooking time: 15 minutes   *Eudroma*
serves: 4-6

1 lb. fresh spinach, cleaned and
    torn
¼ lb. bacon, diced, fried, and
    drained (½ cup)
1 medium onion, thinly sliced
¼ lb. fresh mushrooms, thinly
    sliced

½ cup butter
¼ cup good brandy
¼ cup Cointreau
1 Tblsp. sugar
Juice of one lemon
Salt and pepper, to taste

Melt butter in a flambé pan. Add fried bacon and heat through. Do not burn butter. Add ¼ cup brandy and ignite. When flame dies, add ¼ cup Cointreau and ignite. When flame dies, add sugar, lemon juice and salt and pepper. Cook for approximately 2 minutes. Pour over spinach, mushrooms, and onions while bubbly. Toss through spinach and cover for just a couple of minutes. *Serve.*

DAY BEFORE SALAD

preparation time: 30 minutes   *Maureen Stone*
chilling time: overnight   *Cantamar IV*
serves: 8-10

1 lb. fresh Spinach, cleaned and   1 head Iceberg lettuce, in bite-size
  trimmed   pieces
½ tsp. salt   1 (10 oz.) pkg. frozen peas, thawed
¼ tsp. pepper   1 medium red onion, sliced thin
2 tsp. sugar   2 cups mayonnaise
½ lb. crisp bacon, crumbled   1 cup sour cream
6 hard-cooked eggs, chopped   Sliced Swiss cheese, cut in strips

A large clear salad bowl is best for this. Put spinach on bottom layer.
Sprinkle with half salt, pepper and sugar. Layer bacon, then egg, then lettuce. Sprinkle with remaining salt, pepper, and sugar. Top with peas, then
onion rings. Blend mayonnaise and sour cream. Spread over salad and
top with Swiss cheese strips. Cover and refrigerate overnight. *Serve without tossing.*

ENDIVE, BACON, AND PECAN SALAD

preparation time: 10 minutes   *Kathleen McNally*
cooking time: 10 minutes   *Orka*
serves: 6

1 head butter lettuce, coarsely   1½ tsp. firmly packed brown
  torn (Boston or Bibb)   sugar
½ head curly endive, sliced or   ¼ cup wine, vermouth, or red wine
  coarsely torn   vinegar
¾ cup minced red onion   ¼ tsp. salt or to taste
¾ cup toasted pecan halves (3½   Freshly ground pepper
  oz.)
6 thick bacon slices, cut into ½-
  inch pieces (8 oz.)

The nuts can be toasted ahead of time, but for freshest flavor, this salad
is best put together just before serving.

Combine greens, onion, and pecans in large salad bowl and toss gently.
Cook bacon in large skillet over medium heat until crisp. Drain on paper
towels. Pour off all but ¼ cup fat from skillet. Return to heat, add brown
sugar and stir until dissolved. Blend in vinegar or wine and salt and bring
to a boil. Pour hot dressing over salad. Top with bacon. Season salad
with freshly ground pepper and toss gently. *Serve immediately.*

TUSCAN BREAD SALAD

preparation time: 25 minutes   *Juli Vernon*
chilling time: 20 minutes   *Nordic*
serves: 4

**12 slices stale crusty bread
    (preferably French)
1 garlic clove
1 small cucumber
4 large tomatoes
Salt**

**1 red onion
10 fresh basil leaves
4 Tblsp. olive oil
1 Tblsp. wine vinegar**

Cut the bread into ½ inch thick slices. Peel and halve the garlic clove and rub bread with the cut garlic. Put bread into shallow dish, just cover with cold water and leave for 10 minutes. Meanwhile, wash the cucumber and tomatoes, slice thinly, put into collander and salt. Lift bread out of water and squeeze dry carefully. Arrange in shallow serving dish with cucumbers and tomatoes on top. Slice onions thinly into rings. Arrange over salad with basil leaves. Mix oil, vinegar and season. Pour over and chill for 20 minutes.

The moist garlic flavored bread gives an unusually textured base to this salad. Small slices from a French loaf will give the best appearance.

ITALIAN SALAD

preparation time: 15-20 minutes   *Lori Moreau*
chilling time: 3 hours   *Alberta Rose*
serves: 6

**1 (16 oz.) can Garbanzo beans,
    drained
2 tomatoes, chopped
1 green pepper, chopped**

**¾ cup grated Romano
1 bottle Kraft "Ceasar" dressing
Iceberg lettuce**

Combine beans and vegetables in a medium-sized bowl. Add Ceasar dressing and Romano cheese and toss well. Marinate in refrigerator 3 hours. Line salad bowls with lettuce and spoon marinated vegetables into center.

• *Protect recipe cards from food stains by coating cards with clear shellac.*

GREEK SALAD

preparation time: 10 minutes
marinate: 30 minutes   *Saracen*
serves: 2-4

Cucumber — sliced
Green pepper — sliced in thin
    rings
Red onion — sliced in thin rings
Tomato wedges
Crumbled Feta cheese
Vinaigrette dressing

Dressing:
3 Tblsp. red wine vinegar
¼ cup olive oil
½ tsp. salt
1 tsp. crushed oregano
1 Tblsp. capers
Freshly ground pepper

Combine all vegetables and cheese in salad bowl, prepare dressing and pour over all. Marinate 30 minutes or so before serving.

RADISH-OLIVE SALAD WITH MUSTARD VINAIGRETTE

preparation time: 30 minutes   *Pamela McMichael*
chilling time: 1-2 hours   *Pieces of Eight*
serves: 4

1½ cup sliced radishes
¼ cup pitted Calamata (or black)
    olives, cut into strips

2 Tblsp. chopped scallion with
    tops
4 to 8 Boston lettuce leaves

Combine all and refrigerate, covered. Make vinaigrette.

Mustard Vinaigrette (makes ¼ cup)
3 Tblsp. olive oil
1 Tblsp. white wine vinegar
½ tsp. Dijon-style mustard

¼ tsp. salt
Dash of pepper

Whisk all ingredients together and pour over radish mixture. *Serve on top lettuce leaves.*

ZESTY CAULIFLOWER SALAD

preparation time: 15 minutes
serves: 6-8

Shari Stump
Eyola

1 large head cauliflower,
    separated
Chopped celery
Green onions and chives

Olives
Shrimp (optional)

Mix the following together:

1 cup mayonnaise
5 Tblsp. cider vinegar
Salt and pepper

Dash of Tabasco
Dash of garlic salt

Combine vegetables and mayonnaise mixture. Decorate with cucumbers and cherry tomatoes.

CAULIFLOWER-AVOCADO SALAD

preparation time: 35 minutes
cooking time: 20-25 minutes
chilling time: 3 hours
Serves: 4-6

Shirley Fortune
Maranatha

1 medium head cauliflower
    (approximately 2 lbs.)
½ cup grated Jack cheese

Avocado dip:
2 very ripe avocados, mashed,
    save nut

1 small onion, chopped
1 medium tomato, chopped
1 Tblsp. lemon juice
Salt and pepper, to taste

Trim outer leaves from cauliflower head and slice off stem end so it sits evenly on flat surface. Heat 1" salted water to boiling and add cauliflower, heat to boiling, reduce heat and cook 20 minutes. Drain. Chill 3 hours. Prepare avocado dressing (while cauliflower is cooking) and store, with nut, in tightly sealed container in refrigerator. It will not turn dark! *To serve: place cauliflower on lettuce-lined platter and spread avocado mixture over head. Serve with cheese.*

• *Get more juice from a dried-up lemon by heating for five minutes in water before squeezing.*

CRUNCHY PEA POD SALAD

preparation time: 15 minutes   *Kandy Popkes*
chilling time: 1-2 hours   *Wind Song Oregon*
serves: 8

1 pkg. (6 oz.) thawed pea pods
½ can water chestnuts
1 can (1 lb.) green peas
¼ cup sliced stuffed olives
1 cup thin celery crescents

¼ cup vinegar
¼ cup salad oil
1 Tblsp. soy sauce
½ tsp. paprika
½ cup powdered sugar

Drain all vegetables and combine. Mix remaining ingredients and pour over vegetables. Cover and chill. *Serve on lettuce.*

ALFALFA SPROUTS AND TANGY DRESSING

preparation time: 20 minutes   *Pamela McMichael*
serves: 4   *Pieces of Eight*

1 cup orange sections
2 cups alfalfa sprouts

1 small avocado, peeled and
   sliced
Tangy dressing (recipe follows)

In serving bowl combine sprouts, oranges and avocados. Just before serving, toss with the dressing.

Tangy Dressing:

⅓ cup oil
2 Tblsp. lemon juice
2 Tblsp. pickle relish

1 large clove garlic, minced
½ tsp. sugar
½ tsp. salt

Put all ingredients in a jar and shake until well blended.

This is a delicious dressing.

AVOCADO-CORN SALAD

preparation time: 15 minutes   *Pernilla Stahle*
serves: 6   *Adventure III*

Iceberg lettuce
1 avocado, cut up
1 can corn (or frozen corn)
1 green pepper, cut up

Dressing:

2 Tblsp. olive oil
1 Tblsp. real lemon juice
Salt and black pepper
Salad seasoning

Mix salad ingredients. Pour over dressing. *Serve with or before meal.*

CORN CONFETTI SALAD

preparation time: 10-15 minutes　　　　　　　　*Ann Glenn*
chilling time: refrigerate overnight　　　　　　　*Encore*
serves: 6-8

2 cans whole kernel corn, drained　　1 tsp. salt
4 stalks celery, chopped　　　　　　　1 tsp. lemon pepper
1 medium red onion, chopped　　　　　2 Tblsp. Hellman's mayonnaise
½ green or red bell pepper,　　　　　　2 Tblsp. sour cream (or yogurt)
　　chopped
Juice of one lemon

Mix together all ingredients except dressing (mayonnaise and sour cream
or yogurt). Seal container and let stand in refrigerator overnight. Just before serving, drain again, and add dressing of mayonnaise and sour
cream. Mix well. May be garnished with parsley and paprika.

ITALIAN PASTA SALAD

preparation time: 25 minutes (plus chilling)　　　*Gill Case*
cooking time: 15 minutes　　　　　　　　　　　*M/V Polaris*
serves: 6

8 cups pasta shells　　　　　　　　　2 tomatoes
1 bottle Italian dressing　　　　　　　1 medium onion
1 large cucumber　　　　　　　　　　1 small can black olives
1 small can corn

Cook pasta as directed in salted water. Drain and put in large mixing
bowl. Add enough Italian dressing to coat the shells and leave to cool.
Meanwhile, chop cucumber, tomatoes, onion and olives into small pieces.
Drain corn. Add these to pasta, adding more dressing if necessary. Chill
and serve.

3 BEAN SALAD

preparation time: 5 minutes　　　　　　　　　*Abbie T. Boody*
marinating time: 30 minutes　　　　　　　　　*Ductmate I*
serves: 8

1 can green beans　　　　　　　　　½ red onion
1 can garbanzo beans　　　　　　　　2 oz. sugar
1 can kidney beans
1 cup Good Season salad dressing
　　(prepared)

Mix all together and marinate 30 minutes. *Serve. Will last 5 days in refrigerator.*

GREEN BEAN SALAD PROVENCAL

preparation time: 15 minutes   Jean Thayer
chilling time: 2 hours   Paradise
serves: 6

2 pkgs. frozen whole green beans,   Dressing:
    defrosted   2 cloves garlic, minced
4 ripe tomatoes, peeled, quartered   1 tsp. dried basil
1 cup pitted Nicoise olives   ¼ cup olive oil
Anchovy fillets (garnish)   3 Tblsp. wine vinegar
3 hard-boiled eggs, sliced   Salt and pepper, to taste
    (garnish)

Arrange beans, tomatoes, and olives in a salad bowl. Combine dressing
ingredients and pour over salad. Toss well. Refrigerate 2 hours. Garnish
with anchovies and slices of eggs.

SWISS GREEN BEANS

preparation time: 20 minutes   Kathleen Leddy
cooking time: 10 minutes   Carriba
chilling time: overnight

1½ lbs. green beans   2 large cloves of garlic
⅓ lb. Swiss cheese, thin strips   ½ cup olive oil
½ cup almonds   1 Tblsp. red wine vinegar
½ cup chopped olives   ½ tsp. tarragon
½ cup green pepper, chopped   1 tsp. dill weed
½ cup red pepper, chopped   Dash of pepper and salt
                                     2 tsp. Dijon mustard
Dressing: ½ cup chopped fresh parsley

5 Tblsp. lemon juice

Steam green beans. Drain and cool 10 minutes. Add dressing and cheese.
Cover and refrigerate 2 hours. Add olives and peppers, mix well, cover
and chill overnight. Serve on bed of lettuce topped with chopped almonds.

CHILLED CUCUMBER SALAD

preparation time: 15 minutes   *Lori Moreau*
chilling time: 2-3 hours   *Alberta Rose*
Serves: 6

2 medium cucumbers, sliced   2 Tblsp. tarragon vinegar
1 medium onion, sliced thinly   Dillweed
1 cup sour cream   Salt
½ cup mayonnaise

Combine sour cream, mayonnaise, vinegar and spices. Mix with cucumbers and onion. Chill well.

FRESH MUSHROOM SALAD

preparation time: 15 minutes   *Paula Taylor*
marinating time: 48 hours   *Viking Maiden*

1 lb. fresh mushrooms   12 bacon rashers cooked and
½ pkg. fresh spinach   crisp
                                 4 hard-boiled eggs

Slice mushrooms and divide between 12 salad plates which have been spread with spinach leaves. Cook bacon until crisp and crumble on top of mushrooms. Pour mushroom salad dressing over and top with chopped eggs and sprig of parsley.

Dressing: ¼ cup sugar
1 cup olive oil   1 Tblsp. Durkee salad seasoning
½ cup vinegar (wine)   4 Tblsp. sour cream
2 tsp. dried mustard   ½ lb. fresh mushrooms
2 cloves garlic

Combine all ingredients except the mushrooms and mix in blender. Add mushrooms to dressing mixture and allow to set in 'fridge for a couple of days. When ready to use, put the whole lot into the blender and pour over mushroom salad. Really tasty!

• *Clean pyrex dishes easily by soaking in one part ammonia and two parts hot water.*

MUSHROOM SALAD

preparation time: 15 minutes   *Gill Case*
chilling time: 1 hour   *M/V Polaris*
serves: 6

1 lb. mushrooms, thinly sliced   **2 tsp. Worcestershire sauce**
Large pinch salt   **1 Tblsp. soy sauce**
Large pinch ground black pepper

Put mushrooms in a deep serving dish. Sprinkle with salt, pepper, Worcestershire and soy sauce. Toss to coat well. Leave for one hour in refrigerator. A lot of juice will have drained from mushrooms, this is the sauce and should not be drained off.

COLE SLAW "BRAVO"

preparation time: 25 minutes   *Launa Cable*
serves: 8   *Antiquity*

Medium sized cabbage   **¼ cup water**
1 tsp. salt   **¼ cup Miracle Whip**
¼ cup sugar   **4 carrots**
⅓ cup raisins

Cut middle of cabbage out, hollowing it, but leaving a shell. Shred cabbage and carrots. Mix salt, sugar, water and Miracle Whip. Combine carrots, cabbage, and raisins and toss while slowly adding liquid. Place slaw inside cabbage shell.

CATHEY'S COLE SLAW

preparation time: 25 minutes   *Jan Robinson*
serves: 6-8   *Vanity*

1 medium cabbage   **1 Tblsp. vinegar**
½ lb. carrots   **½ tsp. sugar**
1 small jar pimentos, or radishes, **1 tsp. oregano**
    or apples   **Salt and pepper, to taste**
¼ cup oil
1 Tblsp. mayonnaise, or more to
    taste

Slice the cabbage thinly. Cut the carrots in small cubes and slice the pimentos. Mix the oil, mayonnaise, vinegar, sugar, oregano, salt and pepper. Toss lightly with cabbage mixture. Best eaten within an hour.

MANDARIN CHIX SALAD

preparation time: 15 minutes   *Abbie T. Boody*
serves: 6   *Ductmate I*

1 lb. cooked shredded chicken   2 oz. peanut butter
2 cups shredded lettuce   ½ cup sesame oil
½ cup sesame seeds

Mix all ingredients and serve. Garnish with orange slices.

CHICKEN, GRAPES AND CURRY DRESSING

preparation time: 30 minutes   *Roz Ferneding*
serves: 4-6   *Whisker*

2-4 cups chicken, cut up   Honeydew melon
¾ cup chopped celery
1 can chopped water chestnuts   Curry Dressing:
½-¾ cup slivered almonds, toasted   1 cup mayonnaise
1-2 cups seedless green grapes,   1 cup sugar
   chopped   Worcestershire
Lettuce   1-3 tsp. curry powder

Mix salad ingredients together, saving almonds for topping. Mix dressing
together. Form salad into loaf, place on lettuce leaves, decorate top with
honeydew slices, almonds and/or shredded coconut. *Serve with crackers
or bread.*

GAY'S CIRCLE CARIBBEAN CONCH SALAD

preparation time: 20 minutes   *Gay Thompson*
chilling time: 1-2 hours   *Satori*
serves: 6-8

2 cups alfalfa sprouts   Dressing:
3-4 large conch — cleaned,   ½ cup oil
   cooked and diced   ¼ cup wine vinegar
2 (16 oz.) cans pineapple slices,   ¼ tsp. dill, garlic powder, and dry
   drained   mustard
2 (16 oz.) cans green beans,   ⅛ tsp. salt and pepper
   drained   4 Tblsp. sliced green onions
Paprika   2 Tblsp. chopped parsley

Shake dressing in jar and chill. Line a shallow serving bowl with sprouts.
Place a pineapple ring in the center and the green beans like spokes out-
ward from it. Slice the remaining pineapple rings in halves. Placing the
cut side toward the beans, form a circular pattern out from the center.
Pile the conch in the center; pour on dressing and sprinkle with paprika.

HEART OF PALM AND AVOCADO SALAD

preparation time: 15 minutes   *Silvia Kahn*
chilling time: 1 hour   *Antipodes*
serves: 6

Lettuce (Iceberg or Romaine)   ½ cup wine vinegar
1 can heart of palm   Salt and pepper or Adobo
2 avocados   seasoning
1 big red onion   Fresh garlic, chopped
Fresh parsley and capers,   ⅓ cup mayonnaise
    chopped   3 tsp. ketchup
                                                          Fresh lime
Dressing: 1 egg yolk

1 cup olive oil

Prepare dressing by combining all ingredients and mix well, chill, until ready to serve with salad. Arrange in oval flat salad bowl: lettuce in small pieces, chopped heart of palm, sliced red onion and avocado slices. Decorate nicely with parsley and capers. *Serve with steak dinner.*

ASPARAGUS & PALMS IN RASPBERRY VINAIGRETTE

preparation time: 20 minutes   *Nancy Thorne*
cooking time: 10 minutes   *Gullviva*
serves: 6

18 asparagus spears   ¼ cup whipping cream
½ lb. spinach   2 Tblsp. sherry vinegar
1 small can hearts of palm   Salt and pepper
1 (10 oz.) frozen raspberries,   1 tsp. chopped chives
    drained   1 tsp. chopped parsley
6 Tblsp. olive oil

Trim asparagus to 4½ inch or so. Add asparagus to boiling water and cook uncovered until tender, about 8 minutes. Drain and chill. Halve and chill palm hearts vertically. Clean spinach. Pureé raspberries in blender. Combine olive oil, cream, vinegar, salt, pepper, and add to blender. Blend and adjust flavoring. *To serve, place spinach leaves on salad plates. Arrange 3 spears of asparagus and 3 pieces hearts of palm, decoratively, on each plate. Spoon ribbon of dressing across center. Garnish with chives, parsley and pepper. Accompany with additional vinaigrette.*

SCRUMPTIOUS SALAD

preparation time: 15 minutes   *Gay Thompson*
serves: 8   *Satori*

Lettuce
1 large red onion
1 large tomato
1 large can pickled artichoke
    hearts
1 green pepper (cut in rings)

1 small jar capers
3-4 ozs. blue cheese
1 large can hearts of palm
⅔ cup Zesty Italian dressing
1 Tblsp. dill

Thinly slice onion, tomato, green peppers in rings. Slice hearts of palm crosswise to make several round pieces.

Place lettuce on individual salad plates. Next, place a slice of red onion followed by a ring of bell pepper followed by a tomato slice; surround the tomato with the "coins" of palm and top with a pickled artichoke heart. Sprinkle with capers. Crumble blue cheese into a cup and add Italian dressing and 1 tablespoon dill. *Mix and pour over salad as you serve.*

MARINATED ARTICHOKE HEARTS

preparation time: 5 minutes   *Paula Taylor*
serves:12   *Viking Maiden*

4 cans artichoke hearts, drained
1 head of lettuce, shredded
2 hard-boiled eggs, quartered
2 firm tomatoes, sliced

Dressing:

    1 cup olive oil

½ cup wine vinegar
2 tsp. dried mustard
2 cloves garlic
1 Tblsp. Durkee salad seasoning
¼ cup sugar

Divide artichoke hearts and place on bed of lettuce. Combine all ingredients for dressing, mixing the mustard with the vinegar first, this keeps the mustard from separating once the olive oil is added. Shake or blend and pour over the artichoke hearts. *Garnish with eggs and tomatoes.*

MARINATED HEARTS

preparation time:·5 minutes　　　　　　　　　*Adriane Biggs*
marinating time: 24 hours　　　　　　　　　　*Taza Grande*
serves: 8

2 lb. cans hearts of palms　　　　　**Pimento slices**
3 jars marinated artichoke hearts　　**2 cloves garlic, sliced in half**

Place artichoke hearts with marinade, hearts of palm and garlic in bowl. Cover and let stand for 24 hours. Remove and slice hearts of palm in half. Arrange on serving plate with artichoke hearts. *Garnish with pimento slices.*

TUNISIAN SALAD

preparation time: 20 minutes　　　　　　　　*Juli Vernon*
chilling time: 30 minutes　　　　　　　　　　　*Nordic*
Serves: 4

1½ lbs. cooked waxy potatoes　　**12 small green olives**
8 oz. cooked carrots　　　　　　　**4 Tblsp. olive oil**
3 canned artichoke hearts　　　　**2 Tblsp. lemon juice**
6 oz. cooked peas　　　　　　　　**1 Tblsp. chopped parsley**
2 Tblsp. drained capers　　　　　**¼ tsp. ground coriander**
12 small black olives

Dice the potatoes and carrots. Drain the artichokes and cut into quarters. Put all the vegetables into serving dish with the capers and olives. Whip oil and lemon juice together with the parsley, coriander, and plenty of seasonings (salt and pepper). Pour the dressing and toss lightly. Chill 30 minutes.

ARTIFICIAL CRAB SALAD

preparation time: 15-20 minutes　　　　　　*Ann Doubilet*
serves: 6-8　　　　　　　　　　　　　　　　*Spice of Life*

Frozen artificial crab (comes in 2　　**Mayonnaise — to taste**
lb. pkgs., individually　　　　　　**Salt and pepper — to taste**
wrapped frozen sticks)　　　　　　**Pineapple**
Lemon juice

Unwrap individual sticks. Chop in ½-inch pieces. Break up and mix with mayonnaise, lemon, salt, and pepper. Hollow out pineapple, chop it, and add to crab and put back in shell. *Can serve the crab in a bowl with crackers as hors d'oeuvres. This stuff is wonderful although no one knows quite what it is made of.*

CAPTAIN'S SALAD (Roy's favorite)

preparation time: 30 minutes   Fran Bryson
chilling time: 24 hours   Amoeba
serves: 8

1 head of lettuce, torn
1 (10 oz.) pkg. frozen peas (cooked
    2 or 3 minutes)
1 cup carrots, grated
½ cup (your choice) chopped green
    pepper, radish, or cucumber
½ cup chopped green onions,
    including tops

½ cup chopped hard-cooked eggs
1 cup Spanish peanuts or
    crumbled bacon
1 cup mayonnaise
1 Tblsp. sugar
1 cup grated cheddar cheese

Layer as given in a large flat bottom dish, press with mayonnaise-sugar mixture. Sprinkle with grated cheese, seal tightly and chill for 24 hours (do not peek).

ENSALADA de CAMARON
(Shrimp Salad)

preparation time: 30 minutes   Erica Benjamin
cooking time: 15 minutes   Caribe Monique
serves: 2

2 tsp. fresh lemon juice
¼ cup olive oil
Salt and pepper
1 cup cooked shrimp, coarsely
    chopped
2 Tblsp. minced onion

1 small tomato, cubed
½ small avocado, cubed
2 tsp. parsley
Tortilla cups or bed of lettuce

Whisk oil & juice & seasoning. Add shrimp and onion. Set 15 min. Add tomato and avocado and parsley and toss. Mound in tortilla cups or on lettuce.

Tortilla Cup:
Place tortilla in ladle and hold in place with spoon and fry. These can be held overnight in turned-off oven. *Serve this salad with enchiladas.*

ISLAND SALMON SALAD

preparation time: 20 minutes   *Fran Bryson*
chilling time: several hours   *Amoeba*
serves: 6

2 (8 oz.) cans red salmon   Cucumber Dressing:
2 cups chopped celery   1 cup sour cream
1 medium onion, chopped   2 Tblsp. lemon juice
French dressing   1 Tblsp. prepared mustard
                                   1 cup cucumber (finely chopped,
                                       unpeeled)

Mix ingredients for cucumber dressing in advance and store in the refrigerator. Marinate celery, onions and salmon in a French dressing for several hours until celery has softened. *Serve on a bed of lettuce with cucumber dressing.*

SATORI'S UNDERWAY SHRIMP SALAD

preparation time: 30 minutes   *Gay Thompson*
chilling time: 1-3 hours   *Satori*
serves: 8

4½ oz. small elbow macaroni   4 bell peppers, diced
4 (4 oz.) cans shrimp, drained or 4 green onions, sliced
1 lb. medium peeled shrimp   1 (2½ oz.) jar pimento, diced and
½ cup mayonnaise   drained .
2-3 Tblsp. dijon mustard (¼ cup)   1-2 tsp. cracked black pepper
6 sliced radishes   1 (15 oz.) can asparagus spears,
1 jar marinated artichoke hearts,   drained
    drained, halved   Alfalfa sprouts, optional
4 Tblsp. capers, drained

Precook macaroni as per package directions adding shelled shrimp for the last minute. (Don't cook canned shrimp) Drain, cool and add mayonnaise, mustard (and canned shrimp if that is what you are using). Taste — add extra mustard for zippiness or mayonnaise for creaminess.

Add all ingredients except asparagus and sprouts. Mix and chill.

When you are heeling over and do not feel like cooking, serve this in the cockpit. *Place salad in bowl, decorate with sprouts and asparagus and serve with a round of Brie and Carrs crackers.*

SMOKED OYSTER COCKTAIL

preparation time: 10-15 minutes   Harriet Beberman
chilling time: 30 minutes   Avatar
serves: 4-6

2 cans smoked oysters
Cocktail sauce:
½ cup mayonnaise
1 Tblsp. catsup
½ lemon, squeezed

Dash garlic salt and pepper
Dash salt
Dash hot sauce
¼ tsp. Horseradish

Drain oysters and rinse in cold water, drain, pat dry and chill. Make sauce
by combining all ingredients. *Serve on bed of romaine lettuce over
cracked ice.*

ORIENTAL TUNA SALAD

preparation time: 25 minutes   Barbara Haworth
serves: 6   Ann-Marie II

2 Tblsp. Italian Dressing
2 tsp. curry powder
⅛ tsp. thyme
¼ cup currants or raisins
¾ cup mayonnaise
½ cup water chestnuts, sliced
1 (11 oz.) can mandarin oranges

1 (8 oz.) can pineapple chunks
¼ cup scallions
¼ cup green pepper, chopped
¼ cup celery
2 (6½ oz.) cans white tuna,
    drained
⅓ cup sliced almonds or toasted

Mix dressing, curry, thyme, currants and mayonnaise in a large bowl.
Carefully stir in remaining ingredients except almonds. *Serve on lettuce;
garnish with almonds.*

SALADE NICOISE

preparation time: 30 minutes   Chris Kling
serves: 6   Sunrise

1 head Iceberg lettuce
1 head Romaine
2 cans white tuna
2 cans rolled anchovy fillets with
    capers
3 tomatoes, quartered

3 hard-boiled eggs, quartered
1 can pitted ripe olives
1 purple onion, sliced in rings
Pimientos
1 can white asparagus
Vinaigrette dressing

Arrange torn lettuce leaves in a very large salad bowl to form a bed of
greens. Place crumbled tuna in the center and arrange all other ingredients attractively around the tuna. Pour vinaigrette dressing over all and
serve for lunch with hot croissants.

TUNA MOUSSE

preparation time: 20 minutes
chilling time: several hours
serves: 8

Maureen Stone
Cantamar IV

2 envelopes of unflavored gelatin
½ cup cold water
1 cup boiling water
2 (8 oz.) pkgs. cream cheese,
    softened
2 Tblsp. lemon juice
1 tsp. curry powder, more if
    desired

½ tsp. garlic salt
⅓ cup chopped green onions
2 oz. pimientos, chopped
2 (7 oz.) cans tuna, drained
Sliced almonds (optional)

In large bowl, sprinkle unflavored gelatin over cold water. Let stand 1 minute. Add boiling water and stir till dissolved. With wire whip, blend in cream cheese until smooth. Add all remaining ingredients, blending well. Pour into 5½ cup fish mold. Chill several hours or overnight. Unmold onto large platter. Garnish with parsley and lemon wedges or dress up with sliced olives for eyes, pimiento slice for mouth and sliced almonds for scales.

LOBSTER SALAD "CANTAMAR"

preparation time: 30 minutes
serves: 8

Maureen Stone
Cantamar IV

6 cups cooked, cubed lobster meat
⅔ cup finely chopped onion
⅔ cup finely chopped celery
½ cup chopped green pepper
¼ cup drained capers
¼ cup chopped fresh parsley
1 tsp. tarragon
1½ cup mayonnaise (homemade is
    best)

1 oz. rum
Lime juice, to taste
Salt, to taste
Freshly ground pepper, to taste
Garnish:
Lettuce or Spinach
Tomato wedges
Cucumber spears

Mix all ingredients well. Make a bed of lettuce or fresh spinach on each of eight plates. Spoon lobster salad in center and garnish with tomatoes and cucumber spears.

FRUIT SALAD WITH LIME SAUCE

preparation time: 20 minutes   *Betsi Dwyer*
serves: 6   *Malia*

3 avocados
1 papaya (3 oranges can be
    substituted)
Boston lettuce
Lime Sauce:
1 cup sour cream

⅓ cup sugar
¼ cup lime juice
2 Tblsp. milk
¼ tsp. nutmeg
¼ tsp. cinnamon

Peel and halve avocados and papaya. Rub avocados with some lime juice to prevent them from turning brown. Slice fruit into thin slices and arrange on individual plates covered with Boston lettuce in a fan shape. *Serve with lime sauce.* Lime Sauce: combine all ingredients and whisk until smooth.

MELON SALAD

preparation time: 20 minutes (plus standing)   *Margaret Benjamin*
chilling time: 2-3 hours   *Illusion II*
serves: 6

1 melon
1 lb. tomatoes
1 large cucumber
1 Tblsp. chopped parsley
2 Tblsp. chopped mint or chives

Dressing:
2 Tblsp. wine vinegar
6 Tblsp. salad oil
Salt and pepper
2 tsp. sugar

Peel and chop cucumber; sprinkle with salt, cover with a plate and let stand for 30 minutes, then drain and rinse with cold water. Cut up melon and tomatoes into bite size pieces. Mix with drained cucumbers and add dressing. Chill 2-3 hours. Just before serving, mix in fresh herbs.

WALDORF SALAD

preparation time: 20 minutes   *Cheryl Fleming*
chilling time: 1-2 hours   *Silver Queen of Aspen*
serves: 6-8

4 apples (2 red, 2 golden), chopped
    fine (skin on)
2 stalks celery, chopped fine

½ cup walnuts, chopped medium
Mayonnaise, add as needed
1 tsp. sugar

Add all ingredients together and refrigerate. One charter guest of mine helped me make the salad and decided to add a little curry powder to it for a change.

PEACH SALAD

preparation time: 10 minutes Gay Thompson
serves: 8 Satori

8 peach halves lettuce
8 ozs. cream cheese Paprika
8 ozs. chopped walnuts

Lay a bed of lettuce on individual serving plates. Place a peach half on each — cavity side up. Combine cheese with walnuts and place in center of peach (large round amount). Sprinkle with paprika.

Try apricot halves also.

DIVINE APRICOTS

preparation time: 10 minutes Cheryl Anne Fowler
cooking time: 10 minutes Vanity
serves: 4

1 (16 oz.) can apricot halves 1 tsp. lemon juice
    (water pack) 6 inches stick cinnamon
2 Tblsp. brown sugar Low-fat plain yogurt (optional)

Drain apricot halves, reserving juice; set apricots aside. In small sauce-pan, combine reserved apricot juice, brown sugar, lemon juice, and stick cinnamon. Simmer apricot juice mixture, uncovered, for 5 minutes. Add apricot halves; heat through. Chill thoroughly. Remove stick cinnamon before serving. *Top each serving with a dollop of yogurt if desired.* This makes a nice addition to a lunch.

FRUIT WITH ORANGE CARAMEL SYRUP

preparation time: 15 minutes (plus cooling) Pamela McMichael
cooking time: 15 minutes Pieces of Eight
serves: 6

Assorted seasonal fruit (enough 1 Tblsp. water
    for 6 cups) ¼ cup orange juice
¼ cup sugar ¼ tsp. cardamon

Cut fruit into bite-size pieces and reserve. Place sugar and water in small skillet. Heat till sugar is a dark golden brown, about 8 minutes. Remove from heat and stir in orange juice and cardamon. Return to heat till smooth. Cool. Drizzle cooled syrup over fruit and let stand at room temperature till serving time.

FRUITY RICE SALAD

preparation time: 10 minutes   *Paula Taylor*
serves: 12   *Viking Maiden*

Creamy Pineapple Dressing   **1 cup raisins**
    (recipe below)   **2 apples, chopped**
3 cups cooked white rice   **½ stalk celery, diced**
1 (15¼ oz.) can pineapple chunks, **2 tomatoes, chopped**
    drained (keep liquid for   **1 onion, chopped**
    dressing)   **Salt and pepper**
1 cup chopped nuts

Combine all ingredients in large salad bowl and toss with
Creamy Pineapple Dressing:

Juice from 15¼ oz. can of   **1 clove garlic**
    pineapple   **2 tsp. dried mustard**
1 cup olive oil   **4 Tblsp. sour cream**
½ cup wine vinegar   **Salt and pepper**
3 Tblsp. sugar

Mix all ingredients together, in blender if possible. Toss with Fruity Rice
Salad.

UNDER THE SEA SALAD

preparation time: 5 minutes   *Launa Cable*
cooking time: 10 minutes   *Antiquity*
chilling time: 1 hour
serves: 4-6

1 small pkg. of lemon Jello   **1 cup Miracle Whip**
1 small pkg. of lime Jello   **1 cup cottage cheese**
1 large can of crushed pineapple   **Dash of salt**

Drain pineapple, saving juice. Add water to pineapple juice to make 2
cups. Heat until *almost* boiling. Add Jellos, stir to dissolve. Add pineapple, Miracle Whip, cottage cheese, and salt. Stir. Pour into mold or bowl.
Refrigerate. Stir after 1 hour.

EASY FRUIT SALAD

preparation time: 20 minutes　　　　　　　　　　　*Debbie Olsen*
chilling time: 1-2 hours　　　　　　　　　　　　　*Kingsport*
serves: 8

1 large can of pineapple tidbits
1 large can of mandarin oranges
1 cup seedless grapes
1 cup flaked coconut

2 bananas, sliced
½ cup walnuts, chopped
¾ cup Cool Whip
¼ cup sour cream

Drain all canned fruit and mix together. Combine remaining ingredients. *Chill and serve.*

MANDARIN SALAD

preparation time: 10 minutes　　　　　　　　　*Harriet Beberman*
chilling time: overnight　　　　　　　　　　　　*Avatar*
serves: 6

½ pint sour cream
1 cup minature marshmallows
1 can mandarin oranges, drained
1 (20 oz.) can pineapple chunks,
　　drained

¼ cup coconut, shredded
2 or 3 bananas, sliced

Mix sour cream, marshmallows and coconut. Use wooden spoon and stir in oranges and pineapple. Refrigerate overnight. Just before serving add bananas.

MANDARIN ORANGE SALAD

preparation time: 20 minutes　　　　　　　　　*Louella Holston*
serves: 6-8　　　　　　　　　　　　　　　　　　*Eudroma*

1 head Romaine lettuce
1 medium red onion, sliced
1 can mandarin oranges

¼ cup sliced almonds or water
　chesnuts

Mix all above ingredients in a salad bowl. Toss with celery seed dressing below.

Celery Seed Dressing:
½ cup powdered sugar
¼ cup vinegar
1 tsp. mustard

½ tsp. celery salt
½ tsp. celery seed
½ tsp. paprika
1 cup oil

Mix all of the above well and then slowly add 1 cup oil while beating.

SATORI'S SUPER SAILING FRUIT SALAD

preparation time: 20 minutes *Gay Thompson*
chilling time: 1-2 hours *Satori*
serves: 8

2 apples, cored and diced
2 oranges, seeded and diced
½ pineapple, skinned, cored, diced
½ cup seedless grapes
Melon balls, kiwi fruit, peaches,
   pears, or any fruit available

2 large bananas, sliced
1 8 oz. frozen raspberries,
   thawed, including juice
½ pint sour cream or cottage
   cheese

Mix all fruit including juice from raspberries. Chill. *Serve with sour cream as a dressing. Camembert cheese and Sea Toast or Carrs crackers are a good accompaniment.*

A great lunch to prepare at breakfast when you know you will be beating up the Channel.

RASPBERRY MOLD

preparation time: 15 minutes *Michelle Sutic*
chilling time: 2-4 hours *Harp*
serves: 8

6 oz. red Jello
2 cups boiling water

10 oz. frozen raspberries
1 can (15 oz.) crushed pineapple

Dissolve Jello in boiling water. Drain raspberries and add to hot Jello. Cool slightly and then add pineapple. Pour into mold and refrigerate several hours.

Great side dish for poultry, beef or pork.

TREKKER'S BLUE CHEESE

preparation time: 20 minutes *D. J. Parker*
chilling time: 1 hour *Trekker*

1 small wedge blue cheese
1 cup sour cream
1 Tblsp. mayonnaise

3 Tblsp. milk
¼ tsp. Accent

Crumble blue cheese. Mix with sour cream (one-half at a time). Add mayonnaise. Blend in milk for preferred consistency. Mix in Accent. Chill in covered jar. *Serve with favorite salad greens.*

STRAWBERRY-NUT SALAD

preparation time: 20 minutes
chilling time: 2-3 hours
serves: 8

Debbie Olsen
Kingsport

1 pkg. (6 oz.) strawberry Jello
2 cups boiling water
3 bananas, mashed
1 can (8¾ oz.) crushed pineapple
　with juice

1 pkg. (10 oz.) frozen strawberries
½ cup pecans, chopped
½ cup walnuts, chopped
1 cup sour cream

Dissolve Jello in boiling water. Add all other ingredients except sour cream. Pour ½ of the mixture into a mold and chill until it begins to gel. Spread sour cream on top and pour the rest of Jello mixture into mold. Chill until set. This is really a beautiful salad.

HOT BACON DRESSING FOR SPINACH SALAD

preparation time: 5 minutes
cooking time: 10 minutes
serves: 6

Betsi Dwyer
Malia

½ lb. bacon
4 tsp. brown sugar
½ cup sliced green onions or
　scallions

½ tsp. salt
3 Tblsp. vinegar
¼ tsp. dry mustard
Dash paprika

Dice bacon and fry over medium high heat. Reduce heat and add remaining ingredients. Pour hot dressing over 1½ lbs. spinach mixed with sliced mushrooms and hard-boiled eggs.

ERIC'S SALAD DRESSING

preparation time: 10 minutes
chilling time: 1 hour

Betsi Dwyer
Malia

¼ cup lemon juice
¼ cup cider vinegar
1 tsp. water
2 cloves garlic, finely chopped
3 Tblsp. grated fresh Romano
　cheese

¼ tsp. pepper
1 tsp. honey
1 Tblsp. anchovy paste
¾ cup vegetable oil

Blend all ingredients except oil in blender. Slowly add oil with blender going at low speed. Chill 1 hour before serving. *Great served over hearts of palm and tomato slices on a bed of lettuce. Grate fresh black pepper over top of each salad.*

AVOCADO AND GRAPEFRUIT SALAD

preparation time: 30 minutes   Terry Boudreau
serves: 8   High Barbaree

3 pink grapefruit   **⅓ cup vegetable oil**
1½ Tblsp. lemon juice   **2 avocados**

Peel and section 3 pink grapefruits into a colander set over a bowl, removing the pits and membranes and squeezing the membranes to release the juice. Transfer 1½ Tblsp. of the grapefruit juice to a small bowl and combine it with 1½ Tblsp. lemon juice. Add ⅓ cup vegetable oil, or to taste, in a stream. Whisk the dressing until it is well-combined, and season it with salt and pepper. Halve, peel and pit 2 avocados, cutting them lengthwise into ⅓-inch slices, and toss them gently with the remaining grapefruit juice. Arrange the avocado slices, drained, and the grapefruit sections on salad plates, alternating and overlapping them, and drizzle the dressing over them.

PARSLEY VINAIGRETTE

preparation time: 5 minutes   Paula Taylor
makes: 2 cups   Viking Maiden

1 cup olive oil   **1 Tblsp. Durkee salad seasoning**
½ cup wine vinegar   **¼ cup sugar**
2 tsp. dried mustard   **4 Tblsp. of dried parsley**
2 cloves garlic

Mix all in blender and serve in halved avocados. The colors are lovely and really compliment each other.

POPPY SEED DRESSING

preparation time: 10 minutes   Harriet Beberman
Makes: 3½ cups   Avatar

1½ cup sugar   **3 Tblsp. onion juice**
2 tsp. dry mustard   **2 cups salad oil**
½ tsp. salt   **3 Tblsp. poppy seeds**
¾ cup vinegar

Mix first five items. Add oil slowly and blend thoroughly in blender. Add poppy seeds last. Will keep 2 weeks stored in covered container in refrigerator. *Serve on fruit salads, spinach salad or avocado and grapefruit salad.*

POPPY-SEED DRESSING

preparation time: 10 minutes   Kandy Popkes
chilling time: 1 hour   Wind Song Oregon
Makes: 3½ cups

1 egg
¼ cup granulated sugar
1 Tblsp. prepared Dijon-style
  mustard
⅔ cup red wine vinegar
½ tsp. salt

3 Tblsp. grated fresh yellow onion,
  plus any juice from the
  grating
2 cups corn oil
3 Tblsp. poppy seeds

Combine egg, sugar, grated onion, and juice in the bowl of a food processor fitted with a steel blade. Process for 1 minute. With the motor running, pour in the oil in a slow, steady stream. When all the oil is incorporated, shut off the motor, taste and correct seasoning. Transfer mixture to a bowl, stir in poppy seeds and refrigerate, covered, until ready to use. Makes about 1 quart. *Tart dressings are more stylish now, but this one still seems good to us, particularly on a spinach salad with rings of purple onion, sliced hard-cooked eggs, crumbled bacon and homemade croutons, sauteed in butter with garlic.

ANCHOVY SALAD DRESSING

preparation time: 15 minutes   Phoebe Cole
serves: 6   Barefoot

2-3 anchovy fillets or anchovy
  paste
½ tsp. Dijon mustard

1 tsp. wine vinegar
Ground pepper
½ cup olive oil

Mash anchovies, combine with mustard and vinegar and black pepper. Add olive oil, beat till blended. *Serve at room temperature.* People that "dislike" anchovies, "love" this if they are not told the ingredients! It's actually quite mild.

FRESH FRUIT DRESSING

preparation time: 10 minutes   Jan Robinson
Makes: 3 cups   Vanity

1½ cups sour cream
5 Tblsp. orange juice

6 Tblsp. honey
1½ cups coconut granola

Mix sour cream, orange juice and honey. Sprinkle granola on top of fresh fruits (your choice). Toss with sour cream mixture.

SOUPS AND SAUCES

GAZPACHO

preparation time: 15 minutes   *Patty Dailey*
chilling time: 4 hours   *Western Star*
serves: 6

28 oz. can whole tomatoes
1 (10½ oz.) can chicken broth
2 Tblsp. tomato puree
1 Tblsp. onion
1 Tblsp. green pepper

1 small cucumber
2 cloves garlic
5 Tblsp. olive oil
3 Tblsp. red wine vinegar
½ tsp. salt

Put ingredients half at a time in blender. Chill for at least four hours and serve with an ice cube. Great do ahead soup, keeps well.

GAZPACHO OR SUMMER SPANISH SOUP

preparation time: 30 minutes   *Alison P. Briscoe*
chilling time: 2-3 hours   *Ovation*
serves: 6

10 spring onions
½ large cucumber
2 green peppers
1 large garlic clove
1 slice white bread (thick)
2 (15 oz.) cans tomatoes
5 Tblsp. olive oil

2 cups chicken stock
Salt
Pepper
1 tsp. sugar
3 Tblsp. wine vinegar
1 Tblsp. chopped parsley

Chop onions, cucumber and green peppers. Crush garlic with a pinch of salt. Trim crust of bread and soak in cold water for 2 minutes. Squeeze to remove excess water. Add to chopped vegetables and add tomatoes and oil. Blend half at a time to a coarse puree. Pour into large bowl. Stir in cold stock, salt, pepper, sugar and vinegar. Chill for several hours until ready to serve. *Sprinkle with chopped parsley before serving.*

Jan Robinson from Vanity, suggests serving the chilled soup with garnishes in separate small bowls, such as chopped green pepper, peeled and chopped tomatoes, thinly sliced onions, small pitted black olives and croutons.

CARROT SOUP

preparation time: 15 minutes   *Pernilla Stahle*
cooking time: 35 minutes   *Adventure III*
serves: 6

¾ cup onion, chopped   ½ cup cream
3 cups carrots, sliced   1 Tblsp. butter
4 cups chicken stock   Salt
2 tsp. tomato paste   Pepper
2 Tblsp. rice

Fry onions in butter in a saucepan. Add bouillon, carrots, tomato paste and rice. Simmer for 30 minutes. Mix in a blender. Put back in pan. Add cream, salt, pepper, and butter.

COLD MINTED PEA SOUP

preparation time: 20 minutes   *Juli Vernon*
chilling time: 1-2 hours   *Nordic*
serves: 8

6 cups chicken stock (bouillon)   Salt
1 small onion stuck with 2 cloves   Pepper
1 clove garlic   3 cups heavy cream or yogurt
1 tsp. tarragon   Finely chopped mint (garnish)
3 lbs. peas (frozen)

Cook the onion, garlic, tarragon and peas in the stock until peas are just tender. Remove onion and discard. Puree in blender and season to taste. Add yogurt or cream. Chill and garnish.

TZATZIKI

preparation time: 20 minutes   *Jane Marriott*
chilling time: 1-2 hours   *Jan Pamela II*
serves: 4-6

4 cucumbers   2 Tblsp. wine vinegar
2 cups plain yogurt   Salt to taste
4 cloves garlic, crushed   Fresh dill, snipped

Grate the cucumbers coarsly into a strainer. Let them drain for a while, then squeeze out the excess moisture with your hands. (This may not be very hygienic, but it's efficient!) Mix with the other ingredients and top with chopped dill. Cover tightly and refrigerate for at least an hour before serving to let the flavors develop. *Serve with Pita or coarse textured French bread.*

CUCUMBER SOUP

preparation time: 15 minutes   *Lori Moreau*
chilling time: 1-2 hours   *Alberta Rose*
serves: 6

2 medium cucumbers   **Dash Tabasco**
1½ cups milk   **16 oz. sour cream**
2 cans cream of celery soup

Peel and quarter cucumbers. Combine cucumbers with milk, soup, and Tabasco and process in blender until very smooth. Using a wire whip combine mixture with sour cream. Chill well. *Serve garnished with a thin cucumber slice.*

EASY PUMPKIN SOUP

preparation time: 10 minutes   *Lori Moreau*
cooking time: 30 minutes   *Alberta Rose*
serves: 6

2 cans pumpkin (16 oz.)   **4 oz. butter**
1½ cans evaporated milk   **Nutmeg**
2 cups chicken broth   **Ginger**
1 medium onion, minced

In large saucepan, sauté onion in butter until tender. Add pumpkin, milk and broth and mix well. Add nutmeg and ginger to taste. Simmer 30 minutes.

Debbie Olson on Kingsport, sautes ½ cup of celery with the onion and butter, and adds ½ tsp. white pepper.

FRENCH ONION SOUP

preparation time: 15 minutes   *Chris King*
cooking time: 30 minutes   *Sunrise*
serves: 6

3 Tblsp. butter   **1 Tblsp. Worcestershire sauce**
3 large onions, sliced thin   **1 Tblsp. kitchen bouquet**
2 cans beef broth   **Salad croutons**
½ cup dry white wine   **Parmesan cheese**
1 bay leaf

Sauté sliced onions in melted butter until evenly browned. Add beef broth plus two cans water and all other ingredients except salad croutons and parmesan cheese. Simmer uncovered for 30 minutes. Float croutons in each soup bowl and sprinkle with parmesan cheese before serving.

VEGGIE CALLALOO SOUP

preparation time: 30 minutes   *Paula Taylor*
cooking time: 1-2 hours   *Viking Maiden*
serves: 10

2 onions, sliced	2 Tblsp. chicken stock, dried
2 cloves garlic	⅔ cup of water
2 oz. butter	1 (15 oz.) Cream of Coconut
2 Tblsp. oil	Salt and pepper, to taste
2 (15 oz.) cans spinach	West Indian Hot Sauce, to taste

Sauté onions and garlic in butter and oil until brown, add spinach, chicken stock, water and coconut cream, salt and pepper to taste and add West Indian Hot Sauce, stir until mixed and simmer for one to two hours. The longer this cooks, the better it is. Cool and blend in food processor or blender until smooth, heat and serve.

This soup is great if you have folks with allergies to shell fish on board and tastes every bit as good as the Callaloo soup containing crab. *Serve with hot French bread and serve the West Indian Hot Sauce as a condiment.*

LENTIL-VEGETABLE SOUP

preparation time: 30 minutes   *Jessica Adam*
cooking time: 2½ hours   *Nani Ola*
serves: 8

2 cups lentils	1 clove garlic
4 slices bacon, diced	2½ tsp. salt
½ cup onion, chopped	¼ tsp. pepper
½ cup celery, chopped	½ tsp. oregano
¼ cup carrots, chopped	1 (1 lb.) can tomatoes
3 Tblsp. parsley	2 Tblsp. wine vinegar

Rinse lentils; drain and place in soup kettle. Add 8 cups of water and add remaining ingredients, except tomatoes and vinegar. Cover and simmer 1½ hours. Add tomatoes and vinegar. Simmer, covered, 30 minutes longer. Season to taste. Can be made ahead and refrigerated or frozen.

• *Sprinkle mushrooms with lemon juice before cooking. This will keep them light in color and prevent them from turning dark.*

WHITE WINE ONION SOUP

preparation time: 10 minutes
cooking time: 1 hour
serves: 6

Kathleen McNally
Orka

¼ cup butter (½ stick)
6 cups thinly sliced Bermuda or
 Spanish onions (2½ lbs.)
1 large garlic clove
1 tsp. sugar
2 Tblsp. all purpose flour
5¾ cups chicken broth (preferably
 homemade)
1¼ cups beef broth (preferably
 homemade)

2 cups dry white wine
1 cup water
1½ cups carrots (1x¼-inch)
 julienne
Freshly ground pepper
12 slices Gruyere Garlic toast (see
 recipe)
¼ cup chopped fresh parsley

Melt butter in heavy large saucepan over low heat. Add onion and garlic.
Stir to coat. Cover and cook for 20 minutes, stirring occasionally. Sprinkle with sugar, increase heat to medium and cook uncovered until onion
is a deep amber color, about 30 minutes, stirring frequently. Reduce heat
to low, blend in flour and stir 3 minutes. Add broths, wine and water, increase heat and bring to a boil, stirring constantly. Reduce heat to low,
cover partially and simmer soup 20 minutes. Stir in carrots and simmer
until tender. Season soup with pepper. Ladle into heated bowls, top with
2 slices Gruyere Garlic Toast. Sprinkle with parsley.

CHILLED ZUCCHINI SOUP

preparation time: 20 minutes
cooking time: 20 minutes
serves: 8

Jennifer Royle
Tri-world

1 cup green onions, chopped
1 cup celery, sliced
¼ cup butter or margarine
6 medium-size zucchini, sliced (2
 lbs.)
2 cans (13 ¾ oz.) chicken broth

1 tsp. salt
¼ tsp. basil
¼ tsp. pepper
¼ tsp. oregano
1 cup light cream

Sauté onion and celery in butter. When soft, add zucchini, chicken broth
and seasonings. Cook until zucchini is tender (about 10 minutes). Cool.
Puree in blender until smooth. Add cream. Serve chilled.

PARSNIP, POTATO AND LEEK SOUP

preparation time: 10 minutes   Jennifer Royle
cooking time: 30 minutes   Tri-World
serves: 8

2 large parsnip (12 oz.)   **1 cup whipping cream**
1 large russet potato (8 oz.)   **½ tsp. salt**
2 medium leeks (12 oz.)   **Dash white pepper**
5½ cups chicken broth

Shred parsnips and potato. Slice leeks. Reserve 1 cup green leek slices
for garnish. Place vegetables into saucepan, add broth and cook until
tender, about 30 minutes. Strain mixture, reserving liquid. Puree vegetables until smooth, slowly adding liquid. Salt and pepper to taste. Add
whipping cream. *Serve hot or chilled.*

POTATO SOUP

preparation time: 15 minutes   Gay Thompson
cooking time: 40-50 minutes   Satori
serves: 8

7 Large russet potatoes, peeled,   **1 tsp. salt**
   cubed   **Fresh ground pepper**
6 Green onions, sliced   **1 pint sour cream**
4-6 cups water   **¼ cup butter**
1½ cups milk   **½ lb. chopped bacon, fried crisp**
2 Tblsp. dill weed   **Parsley**

Boil potatoes and onions in salted water until they are tender. Add milk
and dill, salt and pepper to taste. Simmer until potatoes start to fall apart
and soup becomes thickened (20 minutes, about). Add butter and 2 tablespoons sour cream. Heat through. *Serve in bowls with a dollop of sour
cream, bacon bits, a dash of dill and parsley on top. It is simply
wonderful!*

BULLSHOT SOUP

preparation time: 5 minutes
cooking time: 10 minutes   Different Drummer
serves: 4

1 can tomato soup   **Sour cream**
1 can bouillon   **Green onions, minced**
cooking sherry

Combine soup and bouillon with two shots of cooking sherry. Heat or
chill, as preferred. *Serve with dab of sour cream or minced green onions.*

BECCA'S CHEESE SOUP

preparation time: 20 minutes   *Jan Robinson*
cooking time: 45 minutes   *Vanity*
serves: 6-8

¼ cup butter (½ stick)
3 onions, chopped
3 stalks celery, chopped
2 grated carrots
2 (10¾ oz.) cans chicken broth
3 (10¾ oz.) cans cream of potato
    soup

8 oz. grated cheddar cheese
Parsley
Dash of tabasco
8 oz. sour cream
3 Tblsp. sherry

Sauté butter, onion, celery, and carrots. Add broth, soup, cheese, parsley, and tabasco. Simmer covered for 30 minutes. Just before serving add sour cream and simmer for 15 minutes. *Add 3 Tblsp. sherry and serve immediately.*

CREAM OF PEANUT SOUP

preparation time: 15 minutes   *Barbara Haworth*
cooking time: 20 minutes   *Ann-Marie II*
serves: 6-8

1 med. onion, chopped
2 stalks celery, chopped
¼ cup butter
3 Tblsp. flour

2 qts. chicken stock
2 cups peanut butter
1¾ cups cream-light
½ cup peanuts, chopped

Sauté onions and celery in butter. Add flour and stir. Add stock and bring to boil. Remove from heat. Add peanut butter and cream; blend. Return to low heat (do not boil). *Serve in bowls; garnish with peanuts. This soup can be served hot or cold.*

BLACK BEAN SOUP

preparation time: 10 minutes   *Silvia Kahn*
cooking time: 20 minutes   *Antipodes*
serves: 6

2 cans unseasoned black beans
2 tsp. (Goya) adobo — all purpose
    seasoning

1 cup onion, chopped
3 cloves fresh garlic
½ cup olive oil

Heat oil in large deep pan. Fry onions and garlic. Add black beans and adobo with pepper and simmer for 15 minutes. (Do not boil beans.)

BEEF-TOMATO BROTH

preparation time: 10 minutes   *Jean Thayer*
cooking time: 5 minutes   *Paradise*
serves: 4

2½ cups beef broth — canned   ½ tsp. Worcestershire
1¼ cup tomato juice   ⅛ tsp. Tabasco — or to taste
1 onion — stuck with 6 cloves   ⅛ tsp. baking soda
3 Tblsp. sugar

Heat all ingredients for 5 min. or 'till hot. Discard onion. Ladle into bowls
— garnish with thin slice of orange. The orange definitely changes the
taste and is more than a garnish.

Easily doubled — use only 1 onion.

BLACK BEAN SOUP

preparation time: 5 minutes   *Betsi Dwyer*
cooking time: 15 minutes   *Malia*
serves: 6

2 cans condensed black bean soup   ¼ tsp. garlic powder
2 soup cans full light cream   ½ tsp. powdered bay leaf
2 tsp. curry powder   2 lemons
1 tsp. cracked pepper   1½ cups sour cream
1 tsp. salt

Slowly stir soup into an equal measure of cream over the lowest possible
flame. Add the spices and continue heating and stirring until the soup
seems about to boil. Add the juice of one lemon and stir until the soup
returns to the boiling point. Pour boiling soup into chilled cups. Top with
a dollop of sour cream and a paper-thin slice of lemon. Tastes as good, or
better than, homemade, but is so much quicker to prepare!

• *Keep a sauce warm without allowing a skin to form by placing a circle
of buttered or damp waxed paper over it. Sprinkle a sweet custard
sauce with very fine sugar to prevent a skin forming.*

CHINESE SOUP

preparation time: 10 minutes
cooking time: 20 minutes
serves: 6

Guy Howe
Goodbye Charlie

2 (16 oz.) cans chicken broth
¼ cup soy sauce
1 Tblsp. chopped scallions or
   onions
1 (6 oz.) can sliced mushrooms

½ cup diced smoked ham
   (optional)
1 (4oz.) pkg. cellophane noodles,
   cooked
1 pinch black pepper

Empty the cans of broth in a large pot and add 2 cans water. Put on medium heat. Add soy sauce, scallions, mushrooms, ham and black pepper. Bring to a boil. Add noodles, prepared in accordance with directions on package. Simmer for 5 minutes.

SOUP TORTOLA

preparation time: 10 minutes
cooking time: 1 hour
serves: 6

Silvia Kahn
Antipodes

2 chicken bouillon cubes
2 chickens, cut up
6 cups of water
½ cup onion, diced
2 Tblsp. olive oil
Fresh garlic
1 tsp. curry or turmeric

½ tsp. nutmeg
Salt
Pepper
1 oz. rum
6 scallions or shallots, chopped
3 eggs

Boil chickens in 6 cups water and bouillon. When fully cooked, remove from broth and remove skin. Cool. Fry onions in oil and add to broth along with nutmeg, salt, pepper, rum, garlic, and curry. Combine shallots/scallions with eggs. Add slowly to boiling soup and cook 3 minutes. *Serve*

• *For a change, float plain or cheese popcorn on your soup. It looks pretty and it's easier than croutons.*

EGG DROP SOUP

preparation time: 5 minutes
cooking time: 5-10 minutes
serves: 6

Adriane Biggs
Taza Grande

6 cups water
3 Knorr clear chicken broth cubes
2 tsp. soy sauce
1 tsp. sesame oil

3 eggs, beaten with a fork
Chopped scallions (green part
   included)

Bring water to a boil. Add chicken broth cubes, soy sauce and sesame oil. Add eggs, stir gently until eggs set. Pour into soup bowls. Garnish with chopped scallions.

LOBSTER BISQUE

preparation time: 1 hour
cooking time: 45 minutes
serves: 4

Kathy Wagner
Planktos

1 lobster 1½-3 lbs. or more
Chicken bouillon
1 onion, sliced
3 cloves garlic
2-4 stalks celery, chopped
1½ cups sherry, dry

Paprika
Peppercorns
1 bay leaf
4 Tblsp. butter
4 Tblsp. flour
3 cups condensed milk or fresh

Boil lobster 10 minutes. Remove and reduce liquid to 2-3 cups. During reduction add chicken bouillon, celery, onions, peppercorns, bay leaf, garlic. Melt butter over low heat and stir in flour until smooth. Add milk while stirring over low heat. When well blended, add to stock. Stir over low heat.

Remove vegetables with slotted spoon. Chop lobster meat into bite size pieces and return to bisque. Add sherry and simmer 5 minutes. Add dash paprika.

CRAB BISQUE

preparation time: 10 minutes   *Harriet Beberman*
cooking time: 10 minutes   *Avatar*
serves: 4

1 (6½ oz.) can crabmeat
1 (10½ oz.) can cream of tomato
   soup

1 (10½ oz.) can green pea soup
1 cup cream or half and half
½ cup sherry

Clean crabmeat. Place soups, crab and cream in saucepan and heat, stirring constantly until well blended and hot. Just before serving, add sherry gradually. Serve hot with bread sticks. Seafood Bisque: Substitute 1 lb. shrimp or 1 pint oysters for crab.

For a light luncheon, I serve the bisque with a Wilted Spinach Salad, French bread and Melon Balls.

THE BEST FISH CHOWDER

preparation time: 30 minutes   *Louella Holston*
cooking time: 15-20 minutes   *Eudroma*
serves: 6

1 medium onion, coarsely chopped
3 cloves garlic, crushed and
   minced
2 stalks celery, sliced
1 small zucchini, cut into ¼-inch
   chunks
1 large can tomatoes, broken up
Basil

Dill
Arromat
Maggie
Tabasco
1 packet of shrimp boil
2 cups cooked white fish
½ cup white wine

Sauté onion, garlic, celery, and zucchini in olive oil and butter. Add spices, Tabasco and shrimp boil. Add fish; stir in white wine and heat slowly for about 15-20 minutes

• *Keep Saran Wrap in freezer for easy handling.*

CONCH CHOWDER

preparation time: 40 minutes
cooking time: 3 hours
serves: 6

Kandy Popkes
Wind Song Oregon

¼ lb. diced salt pork (bacon)
2 onions, chopped
4-5 cloves garlic, minced
1 green pepper, chopped
2 (1 lb.) cans tomatoes
1 (6 oz.) can tomato paste
2 qts. hot water
2 Tblsp. vinegar
10 bay leaves

1 Tblsp. oregano
1 tsp. basil
1 tsp. poultry seasoning
1½ tsp. salt
½ tsp. pepper
few drops Tabasco
8 large conch (ground)
3-4 medium potatoes
1 (12 oz.) can corn, drained

Fry bacon till crisp. Remove from drippings and cook onion, garlic, and pepper in drippings. Stir in other ingredients *except* conch, potatoes and corn. Simmer 15 minutes. Add conch to hot mixture. Bring to boil, reduce heat and simmer 3 hours or until conch is tender. Last 15 minutes add potatoes and corn. Cook until potatoes are done. Adjust seasonings. (I tend to add a little less water for better flavor.)

SAVORY SALMON CHOWDER

preparation time: 30 minutes
cooking time: 20 minutes
serves: 6

Kandy Popkes
Wind Song Oregon

1 can (15½ oz.) salmon
5 slices bacon, diced
½ cup onion, chopped
½ cup green pepper, chopped
2 cups diced potatoes
1 pkg. (10 oz.) frozen mixed
   vegetables
2½ cups water
2 chicken bouillon cubes

1 can (13 oz.) evaporated milk
1 can (8½ oz.) cream style corn
½ tsp. salt
⅛ tsp. pepper
½ tsp. celery salt
½ tsp. dried dillweed or dill seed
½ tsp. curry powder
¼ cup sherry
1 Tblsp. minced parsley

Drain and flake salmon, reserving liquid. Sauté bacon, onion and green pepper. Add potatoes, mixed vegetables, water and bouillon cubes. Cover and simmer 10-12 minutes or until vegetables are tender. Add salmon and liquid, evaporated milk, corn and seasonings. Heat through. Sprinkle with parsley before serving. Thicken if desired.

BAHAMIAM CONCH CHOWDER

preparation time: 45 minutes   Carol Muller
cooking time: 20-30 minutes   S'oublier
serves: 4-6

2 tenderized fresh conchs, cleaned   Paprika
1 large onion   Salt, oregano, basil
2 Irish potatoes   Pepper
1 Tblsp. butter   1 can evaporated milk
Parsley

Cut tenderized (put through tenderizer twice) conch into small pieces.
Add enough water to cover. Boil 5 minutes. In another saucepan boil
diced onion and potato until tender. Add conch and water mixture to
this. Add salt, pepper, and herbs to taste. Reheat to boiling, add evaporated milk. Turn off flame, add butter, stir, place lid on pot. Let stand 5
minutes. Garnish with freshly chopped parsley and paprika. *Serve with favorite cracker or ½ hot hard roll with cheese of choice melted on top. A
totally acceptable chowder can be made using the canned conch, if fresh
is not available.*

HASTY CHOWDER

preparation time: 10 minutes   Ann Glenn
cooking time: 15-20 minutes   Encore
serves: 8-10

2 cans condensed cheddar cheese   3 small or 2 large cans chunk
  soup   tuna, drained
2 cans stewed tomatoes, liquid   1 pint, or more, milk
  included
2 cans whole kernel corn, liquid
  included

In a large pot, combine contents of all cans and simmer slowly. About 10
minutes before serving, add milk and reheat; do not boil. The amount of
milk is intended to be sure that you have made enough. *Serve with hot
cornbread muffins for a light luncheon.*

OCTOPUS (PULPO) CHOWDER

preparation time: 10 minutes
cooking time: 1-2 hours   Different Drummer
serves: 6

1 octopus, cleaned, diced   16 oz. can stewed tomatoes
3 Tblsp. bacon drippings   water to cover
1 large onion, chopped   salt and pepper
3 stalks celery, diced   4 Tblsp. chopped parsley
3 diced carrots   1 Tblsp. thyme

In a large saucepan, sauté chopped octopus in bacon grease. Add all other vegetables and sauté 5 minutes. Add tomatoes, seasonings, and water to make a thick consistency. Cover and simmer 1-2 hours.

EGG AND LEMON SAUCE

preparation time: 10 minutes   Ann Doubilet
cooking time: 15 minutes   Spice of Life
makes: 1 cup

3 egg yolks   Cayenne
Juice of ½ lemon   1 cup chicken stock
1 tsp. arrowroot or cornstarch   Parsley, chopped
1 tsp. salt

Combine in double boiler, egg yolks, lemon juice, arrowroot or cornstarch, salt and cayenne. Whisk together. Slowly whisk in chicken stock, stirring constantly directly over medium heat until sauce thickens. Do Not Boil! When sauce clings to back of spoon, remove from burner and place back in double boiler over hot, not boiling, water. Add parsley just before serving. *Use over breaded butterfly lamb or vegetables.*

WHITE SAUCE

preparation time: 5 minutes   Candy Glover
cooking time: 5 minutes   Anemone
makes: ¾ cup

2 Tblsp. butter   Dash of white wine
2 Tblsp. flour   Salt
½ cup evaporated milk   Pepper

Melt butter over low heat; slowly adding flour, stirring constantly until smooth. Add milk, wine, salt and pepper. Continue stirring until it thickens, 2-3 minutes.

MORNAY SAUCE

preparation time: 15 minutes
cooking time: 30 minutes
makes: 2½ cups

Terry Boudreau
High Barbaree

1 Tblsp. minced onion
3 Tblsp. butter
¼ cup flour
3 cups scalded milk
¼ tsp. salt

White pepper, to taste
2 Tblsp. grated parmesan cheese
2 Tblsp. Gruyere cheese
2 Tblsp. butter

In a saucepan cook 1 Tblsp. minced onion in 3 Tblsp. butter over moderate heat until it is softened. Stir in ¼ cup flour and cook the roux over low heat, stirring for 3 minutes. Remove the pan from the heat and pour in 3 cups scalded milk, whisking vigorously until the mixture is thick and smooth. Add ¼ teaspoon salt and white pepper to taste and simmer the sauce for 15 minutes. Add 2 Tblsp. freshly grated Parmesan cheese, 2 Tblsp. grated Gruyere cheese and 2 Tblsp. of butter. Cook the sauce over moderate heat until the cheese is melted. Strain the sauce through a fine sieve into bowl and cover the sauce with a buttered round of wax paper to prevent film from forming.

BLENDER HOLLANDAISE

preparation time: 5 minutes
cooking time: 1 minute
makes: 2 cups

Adriane Biggs
Taza Grande

6 egg yolks
2 Tblsp. lemon juice
2 Tblsp. sherry

¼ tsp. cayenne pepper
½ tsp. salt
1 cup butter

Have ready in blender: egg yolks, lemon juice, sherry, cayenne papper, and salt. Heat butter to bubbling stage. Cover blender and turn on high. After about 10 seconds, remove lid and pour the butter over the eggs in a steady stream. By the time the butter is poured in, about 30 seconds, the sauce should be finished. *This sauce is excellent for Eggs Benedict.*

HOLLANDAISE SAUCE

preparation time: 5 minutes   *Katie MacDonald*
cooking time: 5 minutes   *Oklahoma Crude II*
serves: 4

4 egg yolks   **Pinch salt**
Juice from ¼ lemon (approx. 1   **2 Tblsp. water**
**   tsp.)**   **Parsley (optional)**
¼ cup butter or margarine

Whisk over low heat until thick.

BANGALAISE SAUCE

preparation time: 5 minutes   *Abbie T. Boody*
cooking time: 5 minutes   *Ductmate I*
serves: 8

1 cup chutney   **2 sliced bananas**
1 stick butter

Sauté all together. *A great sauce on fish, chicken, etc.*

TERIYAKI SAUCE

preparation time: 5 minutes   *Casey Wood*
makes: 1¼ cups   *Megan Jaye*

½ cup soy sauce   **2 Tblsp. raw sugar or 4 pkgs. of**
½ cup water   **   Sweet 'n' Low**
¼ cup Vermouth   **3 cloves garlic, pressed**
                                         Ginger

Mix all ingredients in saucepan. Slowly bring to boil, stirring constantly. Cool.

I marinate chicken in plastic bag at room temperature, rolling accordingly. Marinate seafood in refrigerator. Shrimp and lobster kabobs are a favorite. If you want to make a Teriyaki sauce to serve over, just reserve some of the marinade and add cornstarch over heat.

SWEET & SOUR MARINADE

preparation time: 10 minutes　　　　　　　　*Katie MacDonald*
makes: 1½-2 cups　　　　　　　　　　　　*Oklahoma Crude II*

Juice from 1 lemon　　　　　　**1 Tblsp. garlic powder**
½ cup barbecue sauce　　　　　**1 Tblsp. minced onion**
½ cup catsup　　　　　　　　　**½ cup frozen orange juice, thawed**
¼ cup Worcestershire sauce

Mix all ingredients and marinate meat for at least 2 hours.

BEER BARBECUE MARINADE

preparation time: 5 minutes　　　　　　　　*Abbie T. Boody*
makes: 3½ cups　　　　　　　　　　　　　*Ductmate I*

1 can warm beer　　　　　**½ cup Dijon mustard**
2 cups brown sugar

Mix all together. Marinate chicken, ribs or steaks.

LAMB MARINADE (for 5 lbs. lamb)

preparation time: 10 minutes　　　　　　　*Roz Ferneding*
marinating time: 1 hour　　　　　　　　　　*Whisker*

6 Tblsp. lemon juice　　　　**¾ tsp. cumin**
6 tsp. orange extract　　　　**Dash of oregano**
1½ tsp. salt

Mix together and marinate at least 1 hour at room temperature.

SAUCE FOR STEAKS

preparation time: 5 minutes　　　　　　　　*Erica Benjamin*
serves: 6　　　　　　　　　　　　　　　*Caribe Monique*

¼ cup olive oil　　　　　　　**2 tsp. paprika**

Mix enough oil and paprika to make a thin paste. Coat one side of steak, put on grill, coat the other side. Makes a great crisp outside and keeps it juicy inside.

MUSTARD DILL SAUCE

preparation time: 10 minutes
makes: 1½ cups   *Saracen*

3 Tblsp. sugar   **2 tsp. vinegar**
3 Tblsp. dried dill   **½ cup oil**
4 Tblsp. Dijon   **Mayonnaise**
2 tsp. dried mustard

Combine all ingredients, except mayonnaise, adding oil gradually and whisking. Will keep for ages in refrigerator. When ready to use, add as much mayonnaise as desired for smoothness and thickness. Great with shrimp, tuna, salmon, artichokes, salads.

LIGHT CHICKEN SAUCE

preparation time: 5 minutes   *Candy Glover*
cooking time: 5 minutes   *Anemone*
makes: 1 cup

1 chicken bouillon cube   **1 Tblsp. cornstarch**
½ cup water   **¼ cup evaporated milk**

Mix cornstarch in cold water. Add other ingredients. Heat until bouillon cube dissolves and sauce is thickened.

SWEET GINGER SAUCE

preparation time: 5 minutes   *Katie MacDonald*
cooking time: 5 minutes   *Oklahoma Crude II*
serves: 4

¼ cup water   **2 Tblsp. brown sugar**
2 Tblsp. cornstarch   **2 Tblsp. soy sauce**
2 tsp. grated gingerroot   **Pinch salt**

Stir over low heat until slightly thickened. *Use as sauce for veal, chicken, etc.*

PAULA'S SPECIAL SEAFOOD SAUCE

preparation time: 10 minutes
makes: 5½ cups

Paula Taylor
Viking Maiden

2 cups mayonnaise
2 cups ketchup
1 cup French's mustard

4 Tblsp. Worcestershire sauce
1 Tblsp. hot sauce
Salt and pepper, to taste

Mix all ingredients and serve with shrimp. Used in Moldy Shrimp and as an accompaniment to Jumbo Shrimp.

ORANGE SAUCE

preparation time: 5 minutes
cooking time: 5 minutes
serves: 6

Katie MacDonald
Oklahoma Crude II

½ cup orange juice
2 Tblsp. cornstarch

2 Tblsp. brown sugar
Gratings from 1 orange

Cook over low heat, stirring constantly for 5 minutes

WHIPPED SPICED BUTTER

preparation time: 10 minutes
makes: 1½ cups

Guy Howe
Goodbye Charlie

1 lb. unsalted butter (room
   temperature)
½ cup whipping cream
2-3 Tblsp. powdered sugar

½ tsp. grated nutmeg
½ tsp. ground cardamon

Use hand beater or electric mixer and whip butter until fluffy, adding cream gradually. Then whip in remaining ingredients. Make fancy designs with it in bowl.

Excellent on toast, nut breads, waffles.

DEVONSHIRE CREAM

preparation time: 5 minutes
serves: 6-8

Roy Olsen
Kingsport

3 oz. cream cheese
2 Tblsp. powdered sugar

3 Tblsp. sour cream
1 tsp. vanilla

Beat cream cheese and sugar, then add sour cream and vanilla. *Serve with Aebleskivers, or as a dessert topping.*

WHIPPED TOPPINGS (for puddings and pies)

preparation time: 15 minutes   *Terry Boudreau*
chilling time: 2 hours   *High Barbaree*
makes: 3 cups

½ cup cold water	pinch of salt
1 Tblsp. lemon juice	¼ cup sugar
⅔ cup skim milk powder	¼ tsp. vanilla

Combine water, lemon juice, skim milk powder and salt; blend well. Beat until mixture stands in firm peaks (about 5 minutes). Gradually beat in sugar, then vanilla. Chill.

Honey Whipped Topping:
Make as whipped topping but use ½ cup honey in place of sugar, omit vanilla. Gradually beat honey into whipped mixture.

Maple Whipped Topping:
Make as whipped topping, but use maple syrup (1 cup) in place of sugar and omit vanilla. Gradually beating syrup into whipped topping.

• *To prevent whipped cream from becoming thin after sitting, add one teaspoon unflavored gelatin before whipping.*

APPETIZERS AND BEVERAGES

CREAMY BEEF DIP

preparation time: 10 minutes
chilling time: 30 minutes
serves: 4-6

Pamela McMichael
Pieces of Eight

1 (5 oz.) jar dried beef (Armour)
1 (8 oz.) pkg. cream cheese,
  softened

¼ cup milk
1 tsp. dill weed
½ tsp. horseradish

Combine dried beef that has been rinsed and chopped with cream cheese, milk, dill weed and horseradish. Chill. *Serve as a dip with fresh vegetables.*

CAPTAIN'S FAVORITE!

preparation time: 10 minutes
chilling time: 30 minutes
serves: 8-10

Pamela McMichael
Pieces of Eight

1 cup sour cream
1 pkg. Knorrs vegetable soup mix
1 cup real mayonnaise

1 pkg. chopped spinach, thawed
  and drained
1 can water chestnuts, chopped
  and drained

Mix well and chill. *Serve on cocktail bread, preferably pumperknickle, as it's less salty.*

CALICO MOLD

preparation time: 20 minutes
chilling time: 4-6 hours
serves: 8

Jessica Adam
Nani Ola

1 (2.5 oz.) envelope unflavored
  gelatin
1 cup cold water
1 cup sour cream
2 cups grated sharp cheddar
  cheese
1 (4 oz.) can diced green chiles

2 Tblsp. chopped pimentos
2 Tblsp. chopped olives
1 Tblsp. chopped onion
Crackers or tortilla chips

In a saucepan, sprinkle gelatin over cold water. Let stand 3-4 minutes until softened. Stir over low heat until gelatin dissolves; set aside. In a bowl combine remaining ingredients. Gradually stir in dissolved gelatin. Put into a 4-cup fluted mold. Refrigerate until firm, 4-6 hours. Invert onto a platter and serve with chips or crackers.

EASY SHRIMP APPETIZER

preparation time: 5 minutes   *Jessica Adams*
serves: 4-6   *Nani Ola*

1 can medium shrimp   **1 bottle Seafood Cocktail sauce**
½ pkg. cream cheese

Drain shrimp. Soak in ice water for a few minutes. Drain. Spread shrimp over cream cheese and cover with cocktail sauce. *Serve with crackers.*

BRANDIED CHEESE

preparation time: 10 minutes   *Maureen Stone*
chilling time: 24 hours   *Cantamar IV*
serves: 4

6 oz. Roquefort cheese   **¼ cup brandy**
¼ cup soft butter

24 hours before serving, blend cheese, butter and brandy. Pack in small crock and refrigerate. *Serve with crackers and apple wedges.*

NEVER-FAIL SHRIMP SPREAD

preparation time: 20 minutes
chilling time: 1-2 hours   *Different Drummer*
serves: 4-6

1 can broken shrimp, drained   **8 oz. cream cheese**
2 large celery ribs, diced   **8 oz. mayonnaise**
1 medium onion, diced   **1 envelope gelatin**
1 can tomato soup

Heat soup. Stir in gelatin until dissolved. Add cream cheese; melt, stirring. Remove from heat. Add mayonnaise, onion, celery and shrimp. Stir well and chill (about an hour). Unmold. *Serve with crackers.*

THE FRENCH CONNECTION

preparation time: 5 minutes   *Bobbie Gibson*
serves: 4   *Gibson Girl*

½ cup mayonnaise   **1 tsp. Beau Monde**
½ cup sour cream   **1 tsp. dill**

Mix all ingredients together and serve with chips or fresh vegetables.

ROY'S JAMAICAN PEPPER DIP

preparation time: 5 minutes   *Debbie Olsen*
chilling time: 2 hours   *Kingsport*
serves: 8

16 oz. sour cream   2 tsp. instant minced onion
1 Tblsp. Jamaican choice pepper   1 tsp. adobo crillo
    sauce (more if you like it real 2 Tblsp. dried parsley
    hot!)

Combine ingredients and chill for two hours. *Serve.*

CHEESY VEGETABLE DIP

preparation time: 10 minutes   *Lori Moreau*
chilling time: 2-3 hours   *Alberta Rose*
serves: 6

1 cup sour cream   ¾ cup grated parmesan cheese
¾ cup mayonnaise   ½ Tblsp. Durkee salad seasoning
1 cup grated mozzarella cheese

Combine all ingredients and chill 2-3 hours. *Serve with raw vegetables.*

HOT BROCCOLI DIP

preparation time: 10 minutes
cooking time: 20 minutes
serves: 6-8

Kandy Popkes
Wind Song Oregon

1 (10 oz.) pkg. frozen chopped
   broccoli
3 celery stalks, chopped fine
½ large onion, chopped fine
1 (4 oz.) can sliced mushrooms,
   drained

1 roll garlic cheese
1 can cream of mushroom soup
4 Tblsp. margarine

Cook broccoli as directed and drain well. Sauté celery and onion in 4 Tblsp. of margarine until clear. Add mushrooms and cheese. Stir over low heat until cheese is melted. Add broccoli and mushroom soup. Mix well. *Serve hot in chafing dish.*

PATTY'S VEGGIE DIP

preparation time: 10 minutes
chilling time: 4 hours
serves: 6

Patty Dailey
Western Star

1 Tblsp. Beau Monde or Bonne
   Femme (blend of monosodium
   glutamate, celery seasoning
   and onion powder)
1 Tblsp. fine herbs

1 tsp. parsley flakes
1 cup real mayonnaise
1 cup sour cream

Blend ingredients and chill 4 hours or overnite. *Serve with cut up vegetables. Watch out, it's habit forming!*

MOM'S GUACAMOLE

preparation time: 15 minutes
serves: 6

Patty Dailey
Western Star

2 medium avocados
1 Tblsp. diced onion
1 medium tomato, diced
1 Tblsp. mayonnaise

1 Tblsp. sour cream
1 Tblsp. lemon juice
1 small clove garlic
Dash tabasco

Mash avocados with a fork. *Add remaining ingredients and serve with tortilla chips.*

SHERRY CHEESE BALL

preparation time: 10 minutes   Michelle Sutic
chilling time: overnight   Harp
serves: 6-8

8 oz. cream cheese — softened   ⅓ cup chutney
1 cup sharp cheddar cheese —   ¼ tsp. curry powder
   grated   2 green onions — finely chopped
2 Tblsp. dry sherry   (tops only)

Blend together cream cheese, cheddar, sherry and curry. Pack into a
small bowl lined with plastic wrap. Chill overnight. Unmold and spread
with chutney. Sprinkle green onion over top. *Serve with crackers.*

GUACAMOLE FIESTA

preparation time: 15 minutes   Shirley Fortune
serves: 6-8   Maranatha

1 lg. ripe avocado   1 crushed garlic clove
2 tsp. lemon juice   2 Tblsp. mayonnaise
½ tsp. chile powder   Salt & pepper

Peel and mash avocado. Add lemon juice, chile powder, garlic, mayonnaise, salt and pepper. This may be prepared ahead and refrigerated with
the seed in the mixture.

Topping:

8 oz. sour cream   4 oz. grated cheddar or Jack
1 jar hot taco sauce   cheese
                                    ½ lb. Tostidos or Doritos

To serve, use a pretty dish like a fluted quiche pan and flatten the avocado mound in the center leaving edges clean. Layer sour cream, then taco
sauce and top with mounds of cheese. Surround the edges with chips.
Run chips toward the center and ENJOY!

SHRIMP DIP

preparation time: 15 minutes   Jean Thayer
chilling time: 24 hours   Paradise
serves: 6

1 (4 oz.) can small shrimp   **¼ tsp. dried dill weed**
4 oz. cream cheese, softened   **2 drops Tabasco**
2 Tblsp. butter, softened   **dash of salt**
2 Tblsp. scallions, minced   **Party rye bread**
2 Tblsp. lemon juice

In a small bowl, mix all ingredients thoroughly (except the bread). Cover
and refrigerate overnight. *To serve, mound shrimp dip on serving platter
and surround with party rye bread.*

CAVIAR PIE

preparation time: 25 minutes   Jean Thayer
chilling time: 30 minutes   Paradise
serves: 6-8

4 hard boiled eggs   **2 Tblsp. fresh parsley, chopped**
4 Tblsp. butter, softened   **1 (4 oz.) jar black caviar**
1 medium onion, finely chopped   **Juice of half a lemon**
1 can flat anchovies   **1 cup sour cream**
1 Tblsp. mayonnaise

Mash eggs, butter and half of chopped onion. Spread in the bottom of a
9" pie plate and refrigerate 30 minutes. Mash anchovies to almost a
paste, add mayonnaise and parsley and spread on top of egg. Mix caviar,
juice of lemon and remaining onion and spread on top of anchovy layer.
Refrigerate until serving time. *To serve, carefully spread sour cream
across top. Serve with plain unsalted crackers and dig deeply to get to all
layers of the pie.*

FROSTED LIVERWORST

preparation time: 15-20 minutes
chilling time: 1-2 hours
serves: 6

Betsi Dwyer
Malia

1 (8 oz.) cream cheese, softened
1 medium onion, diced
¼ tsp. garlic salt
½ tsp. lemon juice

Black pepper to taste
½ tsp. paprika
Parsley
1 medium sized liverworst (or 2 small)

Mix together first five ingredients and frost liverworst. Sprinkle with fresh parsley and paprika. Place on waxed paper and chill. Can also be made ahead and frozen. *Serve with melba toast.*

CHUTNEY AND CREAM CHEESE

preparation time: 2 minutes
serves: 6

Betsi Dwyer
Malia

1 (8 oz.) cream cheese   1 jar chutney

Pour chutney over cream cheese and serve with crackers. Great when you're running short on time. Delicious, too!

Jan Robinson from Vanity, suggests using Pick-A-Pepper Sauce in place of the chutney.

HOUMOS

preparation time: 15 minutes
serves: 10

Paula Taylor
Viking Maiden

2 (15 oz.) cans chick peas, save liquid
3-4 cloves garlic

¾ Tblsp. Tahini oil (sesame)
Olive oil

Put chick peas and liquid (from 1½ cans) into blender and blend. Add garlic and tahini oil, blend until smooth. If too thick, add more chick pea liquid. *Pour into small bowls, top with olive oil and serve with Pita bread triangles.*

BABA GHANOUJ

preparation time: 15-20 minutes   Barbara Tyne
cooking time: 1 hour & 10 minutes   Anodyne
chilling time: 30 minutes or longer
serves: 6

1 medium-large eggplant
1 large garlic clove
6 parsley sprigs
1 small onion, peeled and quartered
2 Tblsp. Tahini (sesame seed paste,
    available at health food stores)

1½-2 Tblsp. lemon juice
1 tsp. salt
6 pita breads

Prick eggplant with tip of knife, place on aluminum foil on a cookie sheet and bake in a 400° oven for 1 hour, or until the eggplant collapses when pressed. Remove from oven, slit open lengthwise and allow to cool. In the bowl of a food processor fitted with the steel blade, add the garlic and parsley and process until finely chopped. Add the onion and process with quick on and off turns until finely chopped. Scrape the eggplant pulp into the bowl, making sure all the skin is removed. Add the remaining ingredients and process until smooth. Chill well and serve with toasted pita triangles.

TOASTED PITA TRIANGLES: Cut pita into triangles and pull apart each side. Place in single layers on jelly roll pans and bake at 450° for about 10 minutes, or until hard, but not yet turning brown. *Serve warm or at room temperature.*

BOURSIN BAREFOOT

preparation time: 5 minutes   Phoebe Cole
chilling time: overnight   Barefoot
serves: 6-8

16 oz. cream cheese, softened
8 oz. sweet butter, unsalted
3 cloves garlic, crushed

½ tsp. oregano
¼ tsp. each — basil, dill, marjoram,
    black pepper

Combine softened cheese and butter, add herbs and garlic, blend well. Refrigerate overnight to blend herbs. Also freezes well.

ARTICHOKE DIP

preparation time: 10 minutes   *Gill Case*
cooking time: 15-20 minutes   *M/V Polaris*
serves: 6

1 can (15 oz.) artichoke hearts   ¾ cup mayonnaise
1 cup grated parmesan cheese
   (fresh)

Chop artichoke hearts and put in a mixing bowl. Add parmesan cheese
and mayonnaise. Blend well. Transfer to an ovenproof serving bowl. Bake
at 350° for 15-20 minutes till golden and bubbly. *Serve with Ritz crackers.*

TUNA AND WATERCRESS TARPENADS

preparation time: 20 minutes   *Kathleen McNally*
chilling time: 2-3 hours   *Orka*
serves: 6

1 (7oz.) can chunk white tuna in oil,   2 cups watercress leaves
   drained   4 green onions (including greens),
½ cup mayonnaise   trimmed and minced
4 oz. cream cheese   3 Tblsp. capers
2 Tblsp. fresh lemon juice   2 Tblsp. chopped fresh mint
¼ tsp. freshly ground pepper

Combine tuna, mayonnaise, cream cheese, lemon juice, and pepper.
Blend until smooth. Add watercress, green onions, capers, and mint.
Blend well, but keep some texture. Chill well! *Serve with crisp crackers
and celery stalks. This is a 'snap' using a blender. However, I chop finely
and mix by hand, and it's still an easy success.*

CHEESE BACON SPREAD

preparation time: 20-30 minutes   *Kathleen McNally*
cooking time: 10 minutes   *Orka*
serves: 4

4 slices bacon   ¼ cup mayonnaise
1 (10oz.) pkg. extra sharp cheddar   1 tsp. Worcestershire sauce
   cheese, shredded (2½ cups)   ⅛ tsp. ground red pepper.
½ cup milk

Cook bacon until crisp, drain well, crumble, set aside. Blend cheddar
cheese and remaining ingredients until smooth. Stir in crumbled bacon. If
not serving right away, cover and refrigerate. *Serve with a variety of
crackers.*

SHRIMP FONDUE

preparation time: 5 minutes   Ann Glenn
cooking time: 3 minutes (microwave)   Encore
serves: 6-8

1 clove garlic, cut   8 oz. Swiss or Jarlsberg cheese,
1 can (10¾ oz.) cream of shrimp   grated
    soup, undiluted   2 Tblsp. sherry

Choose a medium size micro-proof bowl. Rub with the cut garlic. Put un-diluted soup, grated cheese, and sherry into the bowl without stirring. Micro 1 minute, stir, micro 2-3 minutes longer, stir well, and serve at once. *Serve with dippers of hot, cubed French bread and fondue forks.*

HOT CRAB DIP

preparation time: 10 minutes   Betsy Sackett
cooking time: 20 minutes   Yacht Insouciant
serves: 4-6

1 can crabmeat, drained   ¼ tsp. marjoram
1 8 oz. pkg. cream cheese   Dash hot sauce
1 Tblsp. minced onion   ⅛ tsp. garlic salt
2 Tblsp. milk   Salt
½ c. white wine   Pepper

Heat cream cheese and milk until melted and smooth. Add remaining ingredients. Slowly simmer until thick, adjusting the consistency with more wine if necessary. *Serve with sesame crackers.*

BARBARA'S DIP

preparation time: 15 minutes   Barbara Haworth
makes: 1½ cups   Ann-Marie II

¾ cup cottage cheese   1 tsp. chopped parsley
¾ cup mayonnaise   1 Tblsp. grated onions
1 tsp. Beau Monde (Spice Island)   1 tsp. dill weed

Blend cottage cheese and mayonnaise. Add rest of ingredients and mix well. *Serve as dip for fresh vegetables.*

PECAN BLUE CHEESE DIP

preparation time: 10 minutes
chilling time: 30 minutes or longer   *Saracen*
serves: 4-6

3 oz. cream cheese	½ tsp. onion powder
8 oz. blue cheese	1 tsp. dill weed
1 Tblsp. white wine	1 tsp. salt
½ cup chopped pecans	½ pint whipping cream

Blend cheeses, wine and spices. Whip in whipping cream. Add pecans. Chill. *Serve with sliced apples and celery.*

CHILIES WITH CHEESE

preparation time: 20 minutes   *Shirley Fortune*
cooking time: 30 minutes   *Maranatha*
makes: 3 cups

1 small onion, finely chopped	Dash of pepper
2 Tblsp. margarine or butter	½ lb. Monterey Jack cheese or
1 cup drained solid pack tomatoes	cheddar
1 can (4 oz.) peeled green chilies, seeded and chopped	¾ cup half and half
½ teaspoon salt	Corn chips

Cook and stir onion in margarine in 10 inch skillet until tender, about 5 minutes. Stir in tomatoes, chilies, salt and pepper. Simmer uncovered 15 minutes.

Cut cheese into ½ inch cubes; stir into tomato mixture. Stir in half and half when cheese begins to melt. Cook and stir until cheese is melted; cook uncovered 10 minutes. *Serve dip in fondue pot or chafing dish with corn chips.*

SATORI'S CLAM DIP AND CHIPS

preparation time: 10 minutes   *Gay Thompson*
serves: 8   *Satori*

1½ cups sour cream	2 Tblsp. prepared horseradish
1 (6½ oz.) can minced clams, drained	3 Tblsp. dill weed

Mix all ingredients. *Let stand 5 minutes and serve with chips.*

ARTICHOKE DIP

preparation time: 10 minutes
cooking time: 20-30 minutes
serves: 8

Harriet Beberman
Avatar

1 can artichokes, drained
1 cup mayonnaise

1 cup parmesan cheese
1 can crabmeat (optional)

Drain, quarter and mash artichokes. Add mayonnaise and parmesan cheese. Put mixture into ovenproof dish and heat in oven at 350° for 20-30 minutes. *Serve hot with crackers.*

Variation: Substitute 1 (8 oz.) pkg. cream cheese for the mayonnaise

Kandy Popkes also adds ½ of a 4 oz. can green chilies, chopped.

SPICY BLUE DIP

preparation time: 15 minutes
makes: 1½ cups

Roz Ferneding
Whisker

4 cloves garlic
3 tsp. Worcestershire sauce
3 Tblsp. vinegar

1 cup plus 1 Tblsp. olive oil
10 oz. blue cheese

Crush garlic. Break blue cheese with fork and mix well, adding Worcestershire and vinegar. Gradually beat in the oil and beat until smooth. This dip can be made in the blender or by hand. *Great with raw vegetables.*

CLAMS GOURMET

preparation time: 15 minutes
cooking time: 10 minutes
serves: 8

Kathleen McNally
Orka

1 lb. Monterey Jack cheese, shredded
3 (6½ oz.) cans chopped clams, drained
2 Tblsp. finely chopped parsley
2 Tblsp. chopped chives

2 garlic cloves, finely minced
Dash ground red pepper
Dash of freshly ground black pepper
8 slices pumpernickle

Combine cheese, clams, parsley, chives, garlic, pepper, and mix until well blended. For each serving, place bread slices in 4 oz. souffle dish, scallop shells (whatever as long as it is ovenproof). Divide clam mixture among dishes. Broil until golden brown and bubbly. Garnish with parsley sprig and lemon wedges. *Serve immediately.*

CHEESE BALL

preparation time: 10 minutes   *Cheryl Fleming*
serves: 8-10   *Silver Queen of Aspen*

12 oz. cream cheese   **Onion and garlic, chopped fine**
2½ cups grated cheddar cheese   **Chopped walnuts**
½ cup French dressing

Mix, fold into ball. Roll in chopped walnuts. *Serve with crackers and fresh fruit.*

CLAM STUFFED EGGPLANT

preparation time: 30 minutes   *Terry Boudreau*
cooking time: 25 minutes   *High Barbaree*
serves: 6

6 eggplants — 3 inches long   **⅓ cup stale bread crumbs**
4 scallions   **¼ cup minced parsley**
2 garlic cloves minced   **¼ teaspoon thyme**
2 Tblsp. olive oil   **2 Tblsp. white wine**
2 Tblsp. butter   **1 tsp. basil**
1 cup minced clams   **1 cup grated Gruyere cheese**

Halve lengthwise 6 eggplants and scoop out the pulp leaving ¼ inch shells. Sprinkle with salt, invert them on paper towels to drain, and chop the pulp. In a skillet saute 4 scallions including the green tops minced, 2 garlic cloves minced, and the eggplant pulp, in 2 Tblsp. each of butter and olive oil, all for 10 minutes, or until the eggplant is softened. Add 1 cup minced clams, ⅓ cup stale bread crumbs, ¼ cup minced parsley, 1 tsp. basil, ¼ tsp. thyme, and salt and pepper to taste. Toss the mixture with a fork, sprinkle it with 2 Tblsp. dry white wine and remove the pan from the heat. Fill the eggplant shells with the stuffing and divide 1 cup grated Gruyere cheese among them. Arrange the eggplants on a buttered baking sheet and bake them in a preheated moderately hot oven at 375° for 25 minutes, or until the cheese is golden.

• *Sometimes food absorbs too much grease while being fried. Prevent this by adding one teaspoon of vinegar to the fat.*

CLAM PUFF APPETIZERS

preparation time: 20 minutes   *Maureen Stone*
cooking time: 5 minutes   *Cantamar IV*
serves: 12

4 oz. cream cheese
¼ cup melted butter
2 (7½ oz.) cans of minced clams (drained)
¼ tsp. rosemary
1 tsp. Worcestershire sauce

½ tsp. salt
1 tsp. grated lemon peel
24 rounds of thin bread, melba or saltines
¼ cup finely chopped parsley

Combine cheese and butter. Stir until well blended. Add next 5 ingredients. Toast bread rounds on one side, pile clam mixture on toasted side. Place on baking sheet and bake for 5 minutes at 400°, or place 4 inches from broiler and watch carefully. Sprinkle with parsley.

AVOCADO-SHRIMP COCKTAIL

preparation time: 10 minutes   *Pernilla Stahle*
serves: 6   *Adventure III*

3 avocados
2 lbs. medium peeled shrimp, cooked
½ cup mayonnaise
½ cup sour cream

2 Tblsp. Gorgonzola or blue cheese
1 Tblsp. crushed walnuts
1 lemon

Cut avocados in halves, squeeze lemon juice on top, place cooled and cleaned shrimp on top. Mix all the other ingredients together and pour on top.

AVOCADO VINAIGRETTE

preparation time: 15 minutes   *Alison P. Briscoe*
serves: 10   *Ovation*

5 ripe avocados (chilled)
1 cup olive oil
¼ cup wine vinegar
salt

Freshly ground pepper
1 tsp. Dijon mustard
1 tsp. sweet basil
¼ tsp. lemon juice

Halve the avocados, place one half in each bowl. Mix all other ingredients. Pour into avocado halves to fill them. *Serve immediately.*

TABBOULEH

preparation time: 30 minutes   *Paula Taylor*
chilling time: 1-2 hours   *Viking Maiden*
serves: 12

2 cups Bulgar wheat   ½ cup fresh parsley (¼ cup if dried)
3 cups boiling water   ½ cup olive oil
2 tsp. salt   1 cup finely chopped whole spring
4 medium tomatoes, diced   onions
½ cup fresh mint (¼ cup if dried)   ½ cup lemon juice

Cover the Bulgar wheat with boiling water, cover and leave to set for ½
hour or longer. Drain, squeeze any excess water from wheat. Combine
Bulgar wheat with remaining ingredients, chill for an hour or longer.
*Serve on bed of Romaine lettuce and Romaine leaves on the side, use
leaves to scoop up Tabbouleh and munch. Tabbouleh can also be used to
stuff into scooped out tomatoes and large mushrooms. Set the stuffed tomatoes and mushrooms on round platter around a large green pepper,
also containing Tabbouleh. This is pretty, but I prefer the Romaine leaves.*

YUMMY STUFFED MUSHROOMS

preparation time: 35 minutes
cooking time: 5-10 minutes   *Saracen*
serves: 6

3-4 slices bacon, cut in ½ inch   4 oz. cream cheese, softened
   pieces   ½ lb. fresh mushrooms (large)

Fry bacon until crisp. Drain and stir bacon into the cream cheese. Preheat
over to 325°. After washing the mushrooms, remove stems. Mound with
cheese mixture. Place on buttered cookie sheet and bake 5 to 10 minutes. Pop under broiler until browned. Watch carefully.

STUFFED MUSHROOMS

preparation time: 15 minutes   *Betsy Sackett*
cooking time: 15 minutes   *Yacht Insouciant*
serves: 4-6

20-24 medium size mushrooms   1 pkg. Jones bulk sausage

Heat oven to 400°. Clean and remove stems from mushrooms. Fill each
mushroom with a well rounded ball of sausage. Bake in foil lined pan until sausage is cooked through, approx. 15 minutes. *Serve with toothpicks
or as a side dish with eggs.*

MUSHROOM NACHOS

preparation time: 15 minutes   Kathleen McNally
cooking time: 8 minutes   Orka
serves: 6-8

24 large fresh mushrooms, cleaned
  well
4-6 Tblsp. butter
2 oz. pepperoni, halved lengthwise,
  thinly sliced
1-2 jalapeno peppers, finely
  chopped

1½ cup Monterey Jack cheese or
  sharp cheddar
1 small red pepper, roasted, or 2 pimentos
¼ cup minced green onions

Line baking sheet with foil. Clean and remove mushroom stems. (Reserve
for other use.) Brush mushroom caps with melted butter. Arrange on
sheet rounded side down. Preheat broiler. Sit one piece of pepperoni in
each cap. Sprinkle with jalapenos, then cheese. Top with red peppers or
pimentos. Broil 6" from heat source, till cheese melts (can bake also).
Transfer to heated platter, sprinkle with minced onions and serve. Zucchini slices, ¼" thick, can be substituted for mushrooms.

COCONUT CHIPS

preparation time: 30 minutes   Terry Boudreau
cooking time: 40 minutes   High Barbaree
serves: 6-8

1 coconut
2 Tblsp. butter, unsalted
Salt

Choose a coconut without any cracks and containing liquid. With an ice
pick, pierce the eyes of the coconut, drain the liquid and reserve it for another use. Bake the coconut in a preheated oven at 400° for 15 minutes,
break it with a hammer, and remove the flesh from the shell levering it
out carefully with a point of a strong knife. Peel off the brown membrane
with a vegetable peeler; cut the coconut meat into 3 inch lengths and approximately 1/16 inch thick slices. Bake the coconut chips in one layer
on a baking sheet in a preheated oven at 250° for 15 minutes. Dot them
with 2 Tblsp. unsalted butter, cut into bits, and sprinkle them with salt.
Increase the heat to moderately slow 325° and bake the chips for 10 minutes, stirring occasionally or until they are golden, and transfer them with
a slotted spoon to serving bowls.

WON TON BOWS

preparation time: 15 minutes
cooking time: 10 minutes
serves: 8

Gill Case
M/V Polaris

1 pkg. won ton skins
2 jars sweet & sour sauce

Cooking oil

Dip your fingers in a bowl of water and spread a stripe of water down the center of each won ton square. Crinkle the center together to form a bow shape. Twist once to secure. Pour oil about ½" deep into skillet. Fry won tons in single layer until golden brown on each side. Drain on paper towels. *Serve with sweet & sour sauce. Can be reheated in 400° oven for 5-7 minutes. Everyone is always fascinated by these!*

RUM NUTS

preparation time: 5 minutes
cooking time: 20 minutes
serves: 4-6

Betsi Dwyer
Malia

4 Tblsp. butter
2 cups pecan halves
4 Tblsp. dark rum

4 tsp. soy sauce
1 tsp. salt
4 drops tabasco

Melt butter in frying pan, spread nuts in butter. Saute slowly, stirring often, until lightly browned. Stir in rum; simmer 1 minute. Add soy sauce, salt, tabasco and stir. Cool on paper towels.

CHEESE PUFFS

preparation time: 15 minutes
cooking time: 15 minutes
serves: 6-8

Abbie T. Boody
Ductmate I

5 different grated cheeses (1½-2 cups)
1 oz. chopped parsley
1 egg
Filo dough (Greek pastry)

½ stick butter, melted
Garnish:
Chopped lettuce
Sliced pimentos

Mix any five cheeses together and add egg and parsley. Cut 2" wide strips of Filo dough and butter with brush. Add ½ oz. of cheese mixture and roll up like a flag. Butter top and bake 350° for 15 minutes. *Serve on bed of chopped lettuce with sliced pimentos scattered for color. This is a great way to use up all those little bits of left over cheeses.*

CRISPY WON TONS

preparation time: 30 minutes
cooking time: 30 minutes
makes: 48

Jeannie Drinkwine
Vanda

1 lb. shrimp, lobster or crab (cooked and chopped)
½ cup water chestnuts, chopped
1 cup bean sprouts, chopped
⅓ cup green onion, finely sliced

2½ Tblsp. parsley, chopped
⅓ cup soy sauce
4 dozen won ton skins (3-inch square)
Oil for frying

In a medium bowl, combine all ingredients, except won ton skins and oil; mix thoroughly. Place about 1 heaping teaspoonful of filing on each won ton skin. Moisten two edges of won ton skin with water, fold skin over to form a triangle. Press moistened edges together to seal. Then moisten one of the corners from the folded edge and press together with the opposite corner. Fry in hot oil at 375° until golden brown. Drain on paper towels and serve immediately with Sweet & Sour Sauce.

Sweet & Sour Sauce:

¼ cup brown sugar
1 Tblsp. cornstarch

¾ cup pineapple juice
¼ cup cider vinegar

SHRIMP EGG ROLLS

preparation time: 20 minutes
cooking time: 45 minutes
serves: 6

Abbie T. Boody
Ductmate I

20 egg roll wrappers
1 lb. shrimp (peeled)
Bunch scallions
Celery (2 stalks)
1 green pepper
Bean sprouts
1 cup shredded lettuce

½ clove ginger and garlic combined
2 chopped carrots
2 oz. soy sauce
½ cup duck sauce
2 oz. oil
Oil for frying

Chop all ingredients. Mince garlic and ginger. In hot skillet or wok, stir fry ginger and garlic, add scallions, green pepper, carrots, shrimp, and celery. Cook 5 minutes. Add sprouts, lettuce, soy sauce, and duck sauce. Let cool and roll in egg roll wrappers. *Fry and serve.*

PUMPKIN FRITTERS

preparation time: 15 minutes   Barbara Haworth
cooking time: 30 minutes   Ann-Marie II
serves: 6-10

2 cups flour
1 tsp. baking powder
Dash salt
¾ cup brown sugar
3 eggs

½ cup milk
1 tsp. vanilla
2 cups cooked, mashed pumpkin
½ cup cooked raisins — optional
Peanut oil

Sift flour, baking powder, and salt together. Cream sugar and eggs; add milk and vanilla. Stir in pumpkin. Add raisins dredged in flour. Heat oil and drop by tsp. Fry until brown, turn once. Two to three minutes per side. *Serve alone or with sour cream dip.*

FRIED CHEESE APPETIZER

preparation time: 20 minutes (plus cooling)   Nancy Thorne
cooking time: 25-30 minutes   Gullviva
serves: 6

6 Tblsp. butter
⅔ cup flour
2 cups cold milk
¼ cup minced onion
1 bay leaf
½ tsp. salt
Pinch nutmeg
Pinch pepper

2½ cups (10 oz.) Swiss cheese,
  grated
2 egg yolks
Cornstarch
2 egg whites, lightly beaten
Bread crumbs, finely ground
6 Tblsp. butter
Homemade tomato sauce

Grease 9 x 9 inch pan. Melt 6 Tblsp. butter, stir in flour and cook 3 minutes in saucepan. Whisk in milk. Add onion, bay leaf, salt, nutmeg, and pepper and simmer 15 minutes. Blend in cheese, whisk in egg yolks one at a time. Discard bay leaf. Pour mixture into prepared pan, smooth top. Let cool. Refrigerate until well set. Turn mixture out onto work surface. Cut into 18 rectangles. Dust with cornstarch, coat with egg white, then cover with bread crumbs. Heat 6 Tblsp. butter in skillet. Add cheese and fry till golden brown, about 5 minutes. *Serve hot with tomato sauce.*

SURPRISE CHEESE PUFFS

preparation time: 15 minutes
cooking time: 10-15 minutes
serves: 4-6

Barbara Haworth
Ann-Marie II

½ cup cheddar cheese — grated
   (room temp.)
¼ cup margarine or butter (room
   temp.)
½ cup flour
½ tsp. paprika

¼ tsp. salt
24 small pimentos, stuffed olives,
   ripe olives, cocktail onions,
   sausage chunks, hot peppers,
   etc. (anything you like)

In small bowl, blend cheese and butter. Add flour, salt, paprika; mix well. Drain "stuffers", blot with paper towel. Divide dough into 24 pieces. Mold dough around "stuffers" completely covering. Store in plastic bag and chill till used, then place on cookie sheet, bake at 425° 15-20 minutes, if frozen, and 8-10 minutes if chilled. *Serve immediately.*

SEA SPROUTS

preparation time: 15 minutes
serves: 8

Gay Thompson
Satori

1 pkg. smoked salmon, thin sliced
8 ozs. cream cheese
5 slices pumpernickle (dark)

Whole black olives, (pitted)
Alfalfa sprouts

Thickly spread 5 slices of pumpernickle with cream cheese. Next, cover with a slice of salmon. Carefully cut bread into 1 inch squares. Top each square with alfalfa sprouts followed by a whole pitted black olive. Pop the whole bite into your mouth at once — tastes as good as it looks, doesn't it?

SALMON TOAST

preparation time: 20 minutes
serves: 6

Pernilla Stahle
Adventure III

6 slices smoked salmon
6 slices toasted white bread (cut
   edges off)
6 tsp. mayonnaise
6 thin slices lemon

6 small fresh dill sticks
6 leaves iceberg lettuce
Salt and pepper
Butter

Butter the toast. Lay other ingredients on toast, in the following order: lettuce, smoked salmon, mayonnaise, lemon, dill garnish. Sprinkle black pepper and a little salt on top.

SATORI'S SWEET CONCH FRITTERS

preparation time: 20 minutes
cooking time: 30 minutes
serves: 4

Gay Thompson
Satori

1 cup sifted flour
1 Tblsp. sugar
1 tsp. baking powder
1 tsp. salt
2 eggs
½ cup milk

1 tsp. salad oil
½ tsp. vanilla
1 (4½ oz.) can crushed pineapple, drained
1 cup conch, cleaned, cooked and diced

Sift flour with sugar, baking powder and salt. Beat eggs with milk, oil and vanilla. Then, gradually add flour mixture, beating till smooth. Add pineapple and conch. In a skillet or pan, heat about 2" oil. When hot, drop by tablespoons full. Cook 3 minutes and turn once. They will be golden brown. Drain well on paper towels. *Serve with cocktail or curry sauce.*

This is a good dessert fritter recipe — omit the conch and use diced apples or apricots.

CRAB BALLS

preparation time: 20 minutes
cooking time: 15 minutes
makes: 24

Jan Robinson
Vanity

2 (6 oz.) cans crabmeat (drained)
3 eggs
1 Tblsp. butter
1 large onion, chopped
1 tsp. chili powder

1½ cups crushed crackers
Salt and pepper to taste
Dash tabasco
1 Tblsp. Worcestershire sauce

Preheat oven to 400°. Saute onions in butter. Mix crabmeat with one cup of crushed crackers and seasonings. Add 2 beaten eggs and a little milk to moisten if necessary. Roll one teaspoon of mixture into balls. Beat remaining egg and dip balls in the beaten egg and roll in the remaining crushed crackers. Place on cookie sheet and bake for 12-15 minutes at 400°, or you can fry them in hot oil in a skillet. Garnish with parsley and lemon wedges. *Serve with spicy cocktail sauce or tartar sauce.*

CHEESE STUFFED PLANTAINS

preparation time: 30 minutes   *Cheryl Anne Fowler*
cooking time: 20 minutes   *Vanity*
serves: 4-6

6 ripe plantains, unpeeled and   ½ tsp. salt
   halved   2 Tblsp. cornstarch
2 qts. (8 cups) water   ½ lb. cheddar cheese, shredded
1½ Tblsp. salt   Oil for frying
4 Tblsp. butter

Bring to boil water, 1½ Tblsp. salt and halved plantains. Cover and boil rapidly for 20 minutes. Drain plantains. Peel, mash and combine with butter, ½ tsp. salt and cornstarch. Cool mixture. Coat hands with cornstarch and take some mixture, form a well in center and stuff with cheese. Cover with more mixture and shape into ball. Repeat to make 16 balls. Deep-fry at 375° until golden. Remove and drain on paper towels. Any kind of stuffing can be used. Get your imagination going and create away!

CHEESE PUFFS

preparation time: 20 minutes (plus chilling)   *Erica Benjamin*
cooking time: 15 minutes   *Caribe Monique*
serves: 4

4 oz. extra sharp cheddar cheese   1 cup flour
   spread, room temperature   Dash of paprika
½ cup unsalted butter, room
   temperature

Combine cheese & butter, blend in flour. Form balls. Dust with paprika. Refrigerate until chilled. Bake 350° for 15 minutes until brown.

GAY'S MEXI-D'OEUVRES

preparation time: 15 minutes   *Gay Thompson*
cooking time: 12-15 minutes   *Satori*
serves: 6-8

2 pkgs. crescent rolls — or   1 (4 oz.) can green chilies, drained
1 pkg. instant rolled frozen bread   and seeded
¼ lb. bacon, fried and crumbled   4-6 ozs. grated cheddar cheese

Unroll dough and place cheese, chilies and bacon on top. Fold dough around it and pinch edges so filling does not leak out (or roll it back up if it is a bread loaf). Bake according to bread directions on package. *Also, can be served as bread with a meal.*

HAM AND CHEESE ROLLS

preparation time: 30 minutes   Linda Richards
serves: 8   Tao Mu

16 thin slices cooked ham   Salt and pepper to taste
8 oz. cream cheese   Celery for garnish
3-4 Tblsp. hot milk   1 slice pimento — chopped
3-4 dashes Tabasco sauce   1 tsp. finely chopped onion

Combine all ingredients except ham and celery. Spread evenly on each slice of ham, roll and secure with toothpicks. Slice celery thin lengthwise in 3-4 inch pieces. Arrange ham rolls and celery attractively on a serving tray. A mild dill dip using sour cream as a base may also be served with this dish to dip the celery in.

HOT HAM AND CHEESE ROLLS

preparation time: 10 minutes   Adriane Biggs
cooking time: 10 minutes   Taza Grande
serves: 6-8

½ cup Poupon mustard   ½ lb. thin sliced boiled ham
2 Tblsp. horseradish   ½ lb. thin sliced Muenster or
                                              Mozzarella cheese

Preheat oven to 400°. Combine mustard and horseradish. Cut ham and cheese slices into 3 x 4 inch rectangles. Spread ham with mustard and horseradish sauce, top with cheese slices. Roll up, jelly-roll fashion. Secure with toothpicks. Place on greased baking sheet. Bake at 400° for 10 minutes, or until cheese melts.

HAM WRAPS

preparation time: 15 minutes   Katie MacDonald
chilling time: 30 minutes   Oklahoma Crude II
serves: 6

1 lb. sliced ham   ¼ cup minced onion
8 oz. cream cheese   2 Tblsp. chives
4 oz. sour cream   1 tsp. garlic powder

Blend all ingredients together, except ham. Spread mix on ham slices and roll up, securing with toothpicks, cut in thirds for easier serving. Chill for 30 minutes or longer, if possible.

HAM AND ASPARAGUS ROLL-UPS

preparation time: 20 minutes   *Barbara Haworth*
chilling time: 1 hour   *Ann-Marie II*
serves: 6-8

12 slices cooked ham   2 Tblsp. mayonnaise
1 pkg. frozen asparagus spears   ½ tsp. horseradish
1 (3 oz.) pkg. cream cheese

Cook, drain, and chill asparagus. Blend cream cheese, mayonnaise, and horseradish. Spread each slice of ham with 2 Tblsp. of cream cheese mixture. Place one large or two small asparagus spears along edge of each slice of ham. Roll up and chill. *Slice each roll into quarters to serve or leave whole and serve as part of cold plate.*

TARA'S BOURBON DOGS

preparation time: 10 minutes   *Jan Robinson*
cooking time: 15-20 minutes   *Vanity*
serves: 8-10

2 (16 oz.) pkgs. hot dogs or   1 tsp. garlic powder
    chicken frankfurters   ½ tsp. onion powder
2 cups ketchup   1 Tblsp. Worcestershire sauce
2 tsp. mustard, prepared   Dash or two of tabasco
½ cup brown sugar   1½ cups bourbon (or to taste)

Cut hot dogs into bite size pieces. Put in a dutch oven and cover with ketchup, add mustard, brown sugar, onion and garlic powder, worcestershire sauce and tabasco. Heat all together. Add bourbon and simmer another 5 to 10 minutes. *Serve warm.*

JAN'S TANGY DOGS

preparation time: 10 minutes   *Guy Howe*
cooking time: 20 minutes   *Goodbye Charlie*
serves: 4-6

1 (16 oz.) pkg. chicken   6 oz. spicy hot mustard, prepared
    frankfurters
8 oz. grape jelly

Cut each frankfurter into four pieces. Combine frankfurters, jelly and mustard in a medium-size saucepan. Cook over a low heat until "dogs" are heated through, stir frequently. *Serve with toothpicks (no buns!!).*

INDONESIAN SAUTAY

preparation time: 10-15 minutes
marinating time: 1 hour
cooking time: 5-6 minutes
serves: 8-10

Ann Glenn
Encore

2 lbs. cooked cubed meat (beef, lamb or pork)
½ cup chunky peanut butter
½ cup soy sauce
¼ cup lemon juice

¼ cup brown sugar
½ tsp. hot pepper sauce (or to taste)
2 large cloves garlic, pressed

Cube meat to small bite-sized pieces. Combine all other ingredients. Marinate meat cubes for one hour or more. Thread meat onto bamboo skewers and broil 3-6 minutes turning once. *Serve hot. (with napkins!)*

A larger serving of sautay over rice can be a light meal when accompanied by salad and bread.

Any excess marinade will keep in the refrigerator for weeks.

HARRIET'S SWEET-SOUR MEATBALLS

preparation time: 30 minutes
cooking time: 30 minutes
serves: 4-6

Harriet Beberman
Avatar

1 lb. chopped beef
¼ cup uncooked quick cooking rice
1 onion, grated
1 clove garlic or ¼ tsp. garlic salt
½ tsp. salt
1 carrot, grated finely

SAUCE:
¼ cup vinegar
½ cup brown sugar
2 cans tomato sauce
½ cup raisins — optional

Combine first six ingredients. Form into balls the size of walnuts and sauté in hot melted shortening or oil until browned. Drain. Make sauce by combining all ingredients (except raisins) in large saucepan or pot with cover. Heat until bubbly, approx. 15 minutes, over moderate heat. Add meatballs to sauce, cook until beginning to boil. Turn heat down and simmer and add raisins. Stir carefully occasionally to keep from sticking or burning on the bottom of the pot. If mixture becomes too thick, ½ cup of water may be added to the sauce. May be frozen. Defrost, reheat as many as needed.

MEXICAN CEVICHE

preparation time: 20 minutes
chilling time: 6 hours
serves: 8-10

Cheryl Anne Fowler
Vanity

1 lb. fresh or frozen fish fillets,
    your choice
1 cup fresh lime or lemon juice
2 medium tomatoes, peeled,
    seeded and chopped

1 small onion, thinly sliced
2-3 pickled serrano peppers,
    rinsed, seeded, and cut in
    strips
¼ cup olive oil or cooking oil
¼ tsp. dried oregano, crushed

Thaw fish, if frozen. Cut into ½ inch cubes. In non-metal bowl cover cubed fish with lime juice. Cover and chill 4 hours or overnight or till fish is opaque, turning occasionally. Add remaining ingredients, ¾ tsp. salt, and ⅛ tsp. pepper to fish. Toss gently to combine; chill.

SEAFOOD RUMAKI

preparation time: 20 minutes
marinating time: 1 hour
cooking time: 10 minutes
serves: 4

Gill Case
M/V Polaris

1 pkg. bacon
1 can smoked oysters or mussels

Soy Sauce

Cut bacon strips in half. Roll each half around an oyster or mussel and secure with toothpick. Marinade these in soy sauce for about 1 hour. Broil rumaki for about 5 minutes, then turn and broil other side. *Serve hot.*

TACO OLIVES

preparation time: 20 minutes
cooking time: 15 minutes
serves: 8

Gill Case
M/V Polaris

1½ lb. ground beef
1½ pkg. taco seasoning
Stuffed green olives

1 jar Hickory Smoked Barbeque
    Sauce

Mix taco seasoning into uncooked ground beef. Form meat mixture around an olive to make a ¾" ball. Repeat until the meat is used up. Bake meatballs in a 350° oven for 15 minutes. Cool completely. These can be frozen until needed. Just before serving, reheat meatballs in Hickory Smoked Barbeque Sauce or a sauce of your choice. *Serve with cocktail sticks.*

SMOKED TROUT MOUSSE

preparation time: 20 minutes   *Kathleen McNally*
chilling time: 3-4 hours   *Orka*
serves: 6

1 lb. smoked trout, skinned and boned
½ cup sliced green onions
¼-⅓ cup loosely packed fresh dill or dried plus 2 tsp.
Fresh lemon juice

¼ tsp. fresh ground pepper
1 cup whipping cream
Salt (optional)
Additional lemon juice (optional)
Fresh dill sprig (garnish)

Combine trout, green onions, dill, lemon juice and ¼ tsp. pepper in processor or blender. (If you have neither, chop ingredients finely with a French knife and combine vigorously.) With the machine running, slowly pour cream in and blend well. Taste, adjust seasoning with salt, pepper and lemon juice. Transfer to crock. Chill several hours. Garnish with fresh dill sprig. *Serve with Rye Heart toast or various crackers.*

COUNTRY FRENCH PATÉ

preparation time: 30 minutes   *Cheryl Fleming*
cooking time: 1 hour   *Silver Queen of Aspen*
serves: 8-12

1 lb. ground sausage meat
½ lb. chicken liver.
⅛ lb. pork fat (fatback)
1 large onion

2 cloves garlic
½ cup brandy
½ cup sweet sherry
2 bay leaves

Sauté sausage meat with chopped onions and chopped garlic. Sauté chicken liver separately. Cube pork fat into small cubes. After chicken liver is sauteed, chop with knife, then blend in blender until fine. Add brandy and sherry liquid to sausage. Let simmer for a few minutes. Remove from stove. After it has cooled somewhat, place sausage mixture in blender very briefly until mixture has a coarse texture. Add the sausage and liver mixture together and mix well adding salt, pepper, celery seed and other spices you like. Place pork fat on top of pàté in a decorative pattern. Garnish with bay leaves on top and bake in a bread pan for 1 hour at 350°. When the top browns and bubbles, it is finished. This Country French Pàté is best a few days after you have baked it. I make a lot of it and freeze it. One pàté will serve two charters because it is so rich. *It is great for a light lunch served with warm rolls and tomato slices.*

SPINACH PATÉ

preparation time: 15 minutes (plus chilling)   Jean Thayer
cooking time: 30 minutes   Paradise
serves: 6

1 lb. fresh spinach — chopped and squeezed dry — or	1 tsp. salt
	¼ tsp. mace
1 (10 oz.) chopped, frozen spinach, thawed and squeezed dry	¼ tsp. cinnamon
	¼ tsp. thyme
	¾ tsp. dried basil
1 lb. good bulk country style pork sausage	

Mix all ingredients — it's super easy in a food processor! Pack into a small baking dish or loaf pan. Bake for 30 minutes at 375°.

Pour off excess fat. Wrap in foil and weight paté with several cans. When cool, refrigerate — still weighted.

To serve, unmold and slice.

May be a 1st course with cherry tomatoes or served as an hors d'oeuvre.

Good for lunch with super-thin pumpernickle and a homemade mustard along with Westphalian ham and Genoa Salami and potato salad.

SEAFOOD PATÉ

preparation time: 20 minutes   Guy Howe
chilling time: 3 hours   Goodbye Charlie
serves: 4

1 can tuna, drained	Worcestershire, to taste
1 (8 oz.) cream cheese, softened	Tabasco (drop)
2-4 Tblsp. chili sauce	1-2 Tblsp. grated fresh onion
Garlic powder	Garnish:
Lemon juice	Lemon slices, parsley, pimento

Mix well, form into loaf and chill 3 hours. Garnish well with lemon slices, parsley, pimiento. *Serve with crackers. Doesn't taste like tuna! Is also very good when left over grouper or dolphin is used, on the rare occasion of left over fresh fish.*

SAILING CHICKEN WINGS

preparation time: 10 minutes   *Jan Robinson*
cooking time: 55 minutes   *Vanity*
serves: 8

2½ lbs. chicken wings, separated   ¼ cup water
    at joints (discard tips)   ½ cup sherry
1 cup soy sauce   2 tsp. dry mustard
1 cup brown sugar, packed   6 scallions, cut in 1" pieces

Combine all ingredients in a medium to large saucepan. Cover, heat to boiling and simmer 30 minutes. Uncover and simmer 15 minutes longer, stirring frequently. *Serve hot or cold.*

LEMON APPLES

preparation time: 10 minutes   *Katie MacDonald*
serves: 4   *Oklahoma Crude II*

4 red apples   Chopped parsley
2 lemons

Wash, core and slice apples. Mix with juice of two lemons. Garnish with parsley and chill if possible. *Lemon Apples served with Ham Wraps, French bread, assorted cheese and wine, makes a cool light lunch.*

BRIE EN CROUTE

preparation time: 20 minutes   *Shirley Fortune*
cooking time: 20 minutes   *Maranatha*
serves: 6-8

2 (4oz.) cans of Brie or   1 pkg. Pillsbury crescent rolls
    Camembert
Dijon mustard

When opening crescents, divide into 2 sections and pinch seams together tightly. Spread a healthy dollop of mustard on each section and place a round of brie on each. Trim the triangles at each corner and bring the edges gently around the cheese. Turn over onto baking dish. Make sure there is no pastry on the bottom side as it won't cook and it becomes soggy. Decorate the tops with left over pastry. Bake 20 minutes at 350°. Let sit a few minutes before serving. Do *not* substitute puff pastry for crescents.

MELON CON HAMON

preparation time: 10 minutes
serves: 6

Silvia Kahn
Antipodes

1 big honeydew melon
6-12 slices of Prosciutto or Laks
ham

Parsley — garnish

Peel and slice melon. Arrange individually on hors d'oeuvre plates. *Top with ham and parsley for decoration and serve with good white French style bread and sweet butter.*

CREAMY FRESH FRUIT DIP

preparation time: 10 minutes
serves: 6

Shirley Fortune
Maranatha

8 oz. cream cheese
7 oz. jar marshmallow cream

1 Tblsp. orange juice — optional
1 tsp. orange rind, grated —
    optional

Combine softened cream cheese and marshmallow until well blended. *Serve with assorted fresh fruits, such as strawberries, grapes, melon chunks, etc.*

BRIE SOUFFLE

preparation time: 10 minutes (plus standing)
cooking time: 30-35 minutes
serves: 4

Kathleen McNally
Orka

3 Tblsp. butter
3 slices white bread
¾ cup milk
2 eggs
Salt, to taste

Freshly ground pepper
Dash Tabasco
½ lb. Brie cheese, cold, rind
    removed, coarsely grated

Preheat oven to 350°. With ½ Tblsp. butter, grease completely a 1 quart baking dish. Butter one side of each slice of bread with remaining 2½ Tblsp. Cut bread into thirds. Whisk together remaining ingredients, except cheese. Arrange half the bread, buttered side up on the bottom of the dish. Sprinkle evenly with half the brie. Top with rest of bread, then brie. Carefully pour egg mixture over it all, and let stand at room temperature for 30 minutes. Bake 30-35 minutes or until golden brown and bubbly.

This soufflé is fool-proof and easily doubled for larger groups.

BRIE A LA CURLY

preparation time: 5 minutes   Kathleen McNally
cooking time: 7-10 minutes   Orka
serves: 6

1 ripe Brie cheese   1 garlic clove, minced
2 Tblsp. butter, melted   Toasted almond slivers

Preheat oven at 450°. Melt butter, add minced garlic, brush evenly over
entire Brie. Place buttered Brie in Pyrex dish. Top generously with slivered almonds. Bake 7-10 minutes, until Brie is lightly brown but doesn't
start to fall and melt. Transfer to warm serving platter, accompanied with
toasted French bread.

CREW'S DELIGHT

serves: 6 (easily expanded)   Jean Thayer
                                              Paradise

4 guests   1 invitation for dinner ashore
2 crew

Day before or day of planned shoreside meal, call on VHF 16 to make
reservations at a suitable restaurant. *To serve: dinghy ashore with Captain, cook, and 4 guests and have a great night out!*

BANANA BONANZA

serves: 4   Saracen

2 ripe bananas (riper the sweeter)   1 Tblsp. wheat germ
1 egg yolk   1 container plain yogurt
1 tsp vanilla   Crushed ice to fill blender

Combine all ingredients and blend. If more sweetness is needed add honey, not sugar. Other fruits may be used.

BANANA COLADA

serves: 4   Roy Olsen
                                                 Kingsport

½ small can Coco Lopez   6 oz. pineapple juice
1 banana   8 oz. light rum

Fill blender with ice. Combine remaining ingredients and whip until ice is
crushed.

BEACHCOMBER

serves: 1

Cheryl Fleming
Silver Queen of Aspen

¾ oz. Scotch
½ oz. Apricot brandy
2 oz. pineapple juice

1 oz. orange juice
Dash lemon juice
Dash lime juice

Garnish with pineapple slice and paper umbrella. Makes one drink.

COLORADO BULLDOG

serves: 1

Debbie Olsen
Kingsport

3 shots Kahlua
2 shots vodka

Cream
Coke

Fill large glass with ice. Layer vodka, Kahlua, cream and top with small amount of Coke. Delicious and dangerous.

CORAL REEF

serves: 1

Gay Thompson
Satori

Fill a cocktail glass with ice. Pour over 2 jiggers of vodka. Add 2-3 Tblsp. Mr. & Mrs. T gimlet mix and stir. Add 2 thin slices of cucumber, 4 large stuffed green olives (I'm addicted to these) and 4 cocktail onions.

A drink and hors d'oeuvres all at once.

EYOLA'S CURE-ALL BLOODY MARY

serves: 1

Shari Stump
Eyola

Vodka
Clammato
Worcestershire sauce

Horseradish
Salt
Slice of lemon

Fill glass with ice, add amount of vodka you prefer. Fill glass with clammato. Add 3 or 4 dashes of Worcestershire sauce, ½ tsp. horseradish, dash of salt and slice of lemon.

FLEMMISH FIZZ

serves: 8   *Kathleen McNally*
                                                Orka

16 ice cubes
2-10 oz. pkg. frozen raspberries in
   syrup, undrained, pureed and
   strained
1 cup fresh orange juice
1 cup whipping cream

⅔ cup gin
½ cup fresh lemon juice
½ cup raspberry brandy or liquor
2 eggs
4 tsp. sugar

Combine all ingredients in blender and mix until smooth. Divide among chilled glasses and serve immediately. Fizzlet — made without gin or brandy.

Perfect eye opener after late night, or for the sometime late sleepers. A special starter for a super sail.

FRESH ORANGE JUICE MIMOSA

serves: 2   *Kathleen McNally*
                                                Orka

2-4 medium oranges, juiced
¼ cup Cognac

1 cup chilled Champagne or
   sparkling wine

Divide juice and Cognac between two fluted glasses and fill each with ½ cup champagne. *Makes 2 servings, but easily adapted for more.*

GREEN GODDESS

serves: 2   *Katie MacDonald*
                                                Oklahoma Crude II

Midori (or any melon liqueur)
¼ cup evaporated milk
1 shot of Rose's lime

Triple Sec, Cointreau or any other
   orange liqueur
Ice

Mix all ingredients in blender and serve.

GOODBYE GUY

serves: 4   *Guy Howe*
                                                Goodbye Charlie

6 oz. vodka
2 oz. gin

8 oz. Midori (melon liqueur)
16 oz. pineapple juice.

Mix well. *Serve over ice cubes. Goodbye Guy!*

ISLAND SPRITZER

serves: 3-4   *Cheryl Anne Fowler*
                                                          Vanity

1 (12 oz.) can papaya nectar,   **1½ cups sparkling white wine,**
**   chilled**   **   chilled**

Gently stir together papaya nectar and sparkling wine. *Serve at once over ice in tall glasses.*

HIBISCUS

serves: 2   *Kyle Perkins*
                                                      Saracen

2 cups pineapple juice   **1 Tblsp. honey**
1 pint strawberries

Blend half strawberries with 1 cup juice. Repeat remaining half and add honey. Chill thoroughly before serving.

MARVELOUS MILKSHAKE

serves: 4   *Erica Benjamin*
                                                   Caribe Monique

1 (13 oz.) can evaporated milk   **Blender full of ice**
Chocolate syrup

Combine milk, ice and chocolate syrup and blend away. For other flavors use fresh fruit, flavored syrup or vanilla.

MIMOSA

serves: 1   *Katie MacDonald*
                                                 Oklahoma Crude II

Ice   **Orange juice**
Champagne   **Orange slice**
Cointreau

Fill glass with ice. Fill half champagne and half with orange juice. Top with a splash of Cointreau and garnish with a slice of orange.

PONCHE COQUITO

serves: 12

Jan Robinson
Vanity

6 egg yolks
1/2 tsp. cinnamon
1 tsp. vanilla
Pinch of salt

2 cans evaporated milk
2 cans Coco Lopez (cream of
   coconut)
1 can sweetened condensed milk
1/2 bottle white rum

Lightly beat egg yolks with cinnamon, vanilla, salt and 1 can evaporated milk. When well mixed, add the other can of evaporated milk, Coco Lopez, condensed milk and the 1 bottle of rum. Pour into a large jar with an air tight lid. The longer it chills in the refrigerator, the better it gets. *Serve straight, in a chilled glass. Tastes like an eggnog, only better!!*

BUSHWHACKER

preparation time: 2 minutes
serves 1 or many

Ellen Stewart
Stewart Yachts

Equal parts of: Baileys
Kahula
Dark Creme de Cocoa
White Creme de Cocoa

Blend with ice and serve, sprinkle with nutmeg.

Note: to put more "whack" in the "bush" add a shot of vodka!

SHERBERT COOLER

serves 2

Katie MacDonald
Oklahoma Crude II

1/4 cup frozen orange juice syrup
1/4 cup evaporated milk
1 shot Rose's lime juice

Ice
Rum or your favorite liquour
   (optional)

Mix all ingredients in blender and serve.

NOONER

serves: 6

Crew
Vanity

2 cups apricot nectar
2 cups peach nectar
2 cups pear nectar

2 Tblsp. honey
Fresh peaches and apricots,
   optional

Combine juices and honey. If adding fresh peaches and apricots, blend well before adding to juices. *Serve over ice.*

ORCHID

serves: 8

Roz Ferneding
Whisker

4 cups purple grape juice
2 cups lemonade

2 cups pineapple juice

Combine ingredients thoroughly and serve over ice. Delicious.

SAILOR'S COFFEE

serves: 1

D. J. Parker
Trekker

½ cup fresh coffee
1 oz. brandy
1 oz. orange liqueur

Whipped cream
Slice of fresh orange

Place brandy and liqueur in decorative mug. Add coffee to about one inch from top. Place orange slice on top edge of mug. Fill with whipped cream. A drop of orange liquer can be added on top. *Serve this dessert coffee immediately.*

SCREAMING WHAT?

serves: 1

Kathi Strassel
Great Escape

3 parts Bailey's Irish Cream
2 parts vodka

1 part Amaretto

Mix ingredients and shake well with ice. *Strain and serve. Better than a Brandy Alexander!*

STRAWBERRY MARGUERITA

serves: 4

Cheryl Fleming
Silver Queen of Aspen

4½ oz. Tequila
1½ oz. Triple Sec
3 oz. lime juice

1 lb. frozen strawberries
Crushed ice

Mix all ingredients in blender and garnish with lime slices. Great with tacos.

SUBTLE AND SUDDEN

serves: 4

Jan Robinson
Vanity

2 cups chocolate ice cream (or vanilla)
6 oz. Bailey's Irish Cream
2 oz. Kahlua

2 oz. Creme de Cacao
3 oz. light rum
3 oz. dark rum

Fill blender ⅔ with ice. Add remaining ingredients; blend. *Serve in champagne glasses or brandy snifters, sprinkle with nutmeg.*

VANITY PUNCH

serves: 4

Jan Robinson
Vanity

1½ cups pineapple juice
1 cup guava juice

½ cup cream of coconut
1 cup white Cruzan rum

Mix thoroughly. Sample, adjust to mood, mmmmm! More rum. *Serve over ice. Garnish with fresh pineapple and cherry.*

ZOMBIE

serves: 1

Cheryl Fleming
Silver Queen of Aspen

½ oz. light rum
½ oz. gold rum
1 oz. Myers Rum
½ oz. 151 Rum
½ oz. apricot brandy

Dash cherry brandy
Dash grenadine
2 oz. pineapple juice
1 oz. orange juice
½ lemon juice

Serve over crushed ice with orange slice as garnish. Serves one.

SUN RAY FIZZ

serves: 1

Pamela McMichael
Pieces of Eight

1½ oz. gin
2 oz. half & half
2 oz. fresh orange juice
1 oz. Holland House Sweet &
    Sour Sauce
½ tsp. vanilla

1 egg white
2 Tblsp. sugar (or to taste)
¾ cup crushed ice
Garnish: ¼ of fresh orange

Mix above ingredients well in a blender.

TERESA'S COQUITO

Jan Robinson
Vanity

2 cans Coco Lopez (cream of
    coconut)
2 cans evaporated milk
1 tsp. vanilla

¼ tsp. cinnamon
Pinch of salt
1 bottle of white rum

Mix all ingredients well and keep in refrigerator until needed. *Serve a*
small amount in a chilled glass (6 oz.)

TROPICAL COOLER

serves: 4

Cheryl Anne Fowler
Vanity

3 Tblsp. creme de menthe
½ of 6-oz. can frozen limeade
    concentrate

1 pint vanilla ice milk
1 (12 oz.) bottle carbonated water,
    chilled

Combine creme de menthe, limeade concentrate, half of the ice milk, and
half of the carbonated water; stir till blended and pour into 4 glasses.
Place additional scoops of ice milk in glasses; fill with remaining carbonated water. Garnish with mint sprigs.

• *Make decorative ice cubes by placing a sprig of mint or a maraschino*
  cherry in each cube before freezing.

VEGETABLES, PASTA, RICE

CURRIED FRUIT

preparation time: 10 minutes   Ann Glenn
cooking time: 13 minutes (microwave oven)   Encore
serves: 8-10

1 large can each: peach halves,	**½ cup toasted almonds**
pear halves, pineapple	**⅓ cup butter**
chunks, maraschino cherries	**½ tsp. curry powder (or more)**
½-1 cup raisins	**¾ cup brown sugar**

Use your prettiest shallow glass baking dish. Arrange drained fruit in an attractive pattern, placing cherries inside pear halves, and sprinkling raisins over all. Combine butter, curry powder and sugar in a glass measuring cup and microwave 2-3 minutes to caramelize. (Watch that it doesn't boil over!) Pour and scrape caramel topping over the fruit. Microwave about 10 minutes total, turning the dish several times. Garnish with the toasted almonds. *Serve hot, as a side dish in place of the usual green vegetables.*

Comment: When Curried Fruit is on the menu, I frequently try to slip past without any additional sweets, feeling I've done enough damage to the calorie count. However, on one occasion, this resulted in a guest saying seriously, with hurt and reproach in his voice, "Did we do a bad thing? Are you mad at us, Ann, that we don't get dessert tonight?" If you suspect your guests may feel slighted, Raspberry Cloud Pie is one of the lightest desserts I know, see Index.

RATATOUILLE

preparation time: 15-20 minutes   Adriane Biggs
cooking time: 1 hour   Taza Grande
serves: 8

1 eggplant, peeled and chopped	**2 tsp. each: basil, oregano**
2 large onions, chopped	**¼ cup parsley**
4 cloves garlic, crushed	**2 tsp. Worcestershire sauce**
½ lb. mushrooms, sliced	**Dash Vermouth**
2 medium zucchini, sliced	**Dash Tabasco**
2 green peppers, chopped and	**¼ cup olive oil**
seeded	
2 lb. can tomatoes and juice	

In large saucepan, saute onion, green pepper, garlic and mushrooms in olive oil, until onions are translucent. Add remaining ingredients, cover and simmer for 1 hour, stirring occasionally. Best when cooked ahead of time, as spices have a chance to mature.

STIR FRIED MIXED VEGETABLES

preparation time: 30 minutes   *Gill Case*
cooking time: 5-8 minutes   *M/V Polaris*
serves: 8

3½ Tblsp. olive oil
1 onion, thinly sliced
3 cloves garlic, crushed
1½ tsp. salt
2 green peppers, cored, seeded,
    sliced
1 red pepper, cored, seeded, sliced
2 large carrots, thinly sliced

½ lb. bean sprouts
Small stalk chinese cabbage,
    chopped
3 stalks celery, chopped
2 Tblsp. soy sauce
2 Tblsp. chicken stock
1½ tsp. sugar

Heat oil in a large skillet over moderate heat. Add onion, garlic and salt and stir-fry for 30 seconds. Add all other vegetables except bean sprouts and toss till well coated with oil. Sprinkle in sugar, add soy sauce and chicken stock. Stir-fry 1 minutes. Add bean sprouts and stir-fry till just cooked. *Serve immediately.*

STEAMED VEGETABLES IN MALT WINE DRESSING

preparation time: 10 minutes   *Pamela McMichael*
cooking time: 10 minutes   *Pieces of Eight*
serves: 4

2 carrots, sliced
½ lb. green beans, cut
2 small zucchini, cut into ¼"
    diagonals
2 Tblsp. olive oil
2 Tblsp. malt vinegar
1 Tblsp. white wine vinegar

½ tsp. salt
¼ tsp. pepper
½ tsp. each: thyme, oregano, and
    rosemary
1 small Boston lettuce — garnish
2 Tblsp. chopped red onion —
    garnish

Steam all vegetables till crisp-tender. Combine olive oil, malt and wine vinegars, herbs, salt and pepper. Whisk and pour over vegetables. Coat evenly. Let stand uncovered at room temperature until serving time. *Serve on lettuce leaves and garnish with red onion.*

• *Vegetables can become limp during storage. Freshen by cleaning them and then submerge in cold water and ice cubes.*

ARTICHOKES WITH LEMON BUTTER

preparation time: 15 minutes
cooking time: 45 minutes to 1 hour
serves: 6

Terry Boudreau
High Barbaree

6 artichokes
1 sliced onion or 1 clove garlic
2 celery ribs with leaves

1½ Tblsp. lemon juice or vinegar
2 Tblsp. salad oil

Cut off stems of the artichokes. Pull off tough bottom row of leaves and cut off the top (½ inch). With scissors clip the tops of the remaining leaves to remove the sharp point. To avoid discoloration, dip cut parts in lemon juice. Place the artichokes on a trivet, upright, with 2 inches of water beneath. Add to water the onion or garlic, celery, lemon juice or vinegar and salad oil. Cook covered for 45 minutes to 1 hour until tender. *Drain them and serve them hot with lemon butter (2 Tblsp. melted butter and ½ tsp. lemon juice per person) in a side dish, one for each person.*

ASPARAGUS WITH NUTMEG BUTTER

preparation time: 10 minutes
cooking time: 10 minutes
serves: 6-8

Chris Kling
Sunrise

3 boxes frozen asparagus spears
   or fresh asparagus, if
   available
3 Tblsp. butter

2 Tblsp. fresh lemon juice
Grated fresh nutmeg

Add asparagus to ½ cup boiling salted water and cook for 8 minutes. Melt butter in small saucepan and add lemon juice. Pour over asparagus in serving bowl and sprinkle with freshly grated nutmeg. Garnish with lemon slices.

GREEN BEANS IN SOY-BUTTER SAUCE

preparation time: 5 minutes
cooking time: 7 minutes
serves: 4

Pamela McMichael
Pieces of Eight

1 lb. beans, trimmed
2 Tblsp. soy sauce

2 Tblsp. unsalted butter

Steam green beans until crisp-tender (about 7 minutes). Meanwhile, combine soy sauce and butter in small saucepan; stir over low heat until butter is melted and mixture is hot. Drain beans and pour sauce over them; toss.

HARICOTS VERTS PROVENCALE

preparation: 10 minutes   *Pernilla Stahle*
cooking time: 10 minutes   *Adventure III*
serves: 6

2 (10 oz.) pkgs. French style green   1 tsp. oregano
   beans, frozen or canned   Butter
1 big onion, finely chopped   Salt
2-3 garlic cloves, crushed   Pepper
3 sliced tomatoes (or canned)   Parmesan cheese
Mayonnaise

Prepare beans according to instruction on package. Fry garlic and onion in butter, add the tomatoes and spices. Add drained green beans. Let fry 5 minutes on low heat. Sprinkle parmesan on top or dot mayonnaise. Excellent dish!

VANITY HOT BEANS

preparation time: 15 minutes   *Jan Robinson*
cooking time: 10-15 minutes (microwave)   *Vanity*
serves: 8

4 slices bacon   1 medium onion, sliced
⅓ cup sugar   1 lb. can cut green beans, drained
1 Tblsp. cornstarch   1 lb. can cut wax beans, drained
½ tsp. pepper   1 lb. can red kidney beans,
½ cup vinegar   drained

In a 2-quart casserole, arrange bacon in single layer. Cover with paper towel and cook 2½-3 minutes, or until crisp. Remove paper towel and bacon, leaving drippings in casserole. To drippings, add sugar, cornstarch and pepper; blend well. Stir in vinegar, then remaining ingredients, mixing well. Cook, covered, 10 minutes, or until sauce has boiled and thickened slightly, stirring occasionally. Crumble bacon over top.

SIMPLY BROCCOLI

preparation time: 10 minutes
cooking time: 8 minutes
serves: 4

Barbara Tyne
Anodyne

1 very large bunch broccoli
1 large clove garlic
⅓ cup freshly grated parmesan

4 Tblsp. butter
Salt, to taste
Pepper, to taste

Cut broccoli into flowerettes, peel stems and cut into ¾ " pieces. Steam broccoli about 4 minutes (may be done ahead and refrigerated at this point). Just before serving, toss broccoli with butter and garlic over high heat and cook just until heated through. Add cheese, remove from heat, toss to coat and serve immediately.

BROCCOLI IN LEMON SAUCE

preparation time: 20 minutes
cooking time: 25 minutes
serves: 4-6

Jessica Adam
Nani Ola

1½ lbs. broccoli
4 Tblsp. butter
2 Tblsp. flour
1 cup milk

2 tsp. grated lemon peel
2 tsp. lemon juice
½ tsp. ginger
½ tsp. salt

Steam cut-up broccoli 15 minutes (for frozen follow directions). Melt butter in small frying pan and gradually whisk in flour slowly, add milk, stirring constantly, to avoid lumps. Gradually, add lemon peel and juice, then add ginger and salt. Stir until well-blended. Spoon over broccoli and garnish with lemon slices.

BROCCOLI CASSEROLE

preparation time: 10 minutes
cooking time: 45 minutes
serves: 8

Kathleen McNally
Orka

2 boxes chopped broccoli (cook as directed)
1 can cream of mushroom soup
1 cup grated American cheese

1 cup mayonnaise
2 eggs
Dehydrated onion flakes, to taste

Preheat oven to 375°. Grease large casserole. Cook broccoli. Combine all ingredients in large bowl, mix well. Place in greased casserole, bake 45 minutes. Cover with crushed Ritz crackers (last 10 minutes). Wonderful side dish.

BROCCOLI AND MUSHROOMS

preparation time: 10 minutes

cooking time: 15 minutes

serves: 6

Terry Boudreau

High Barbaree

2 lbs. broccoli
½ lb. mushrooms
6 Tblsp. butter

½ cup lemon juice
Salt and pepper, to taste

Trim and separate the 2 lbs. broccoli into florets, reserving the stems for another use, and arrange it in a vegetable steamer. Set the steamer in a large saucepan over 1-inch boiling salted water and steam the broccoli, partially covered for 8-10 minutes, or until it is just tender. In a small saucepan combine ½ pound mushrooms, trimmed and cut into ½-inch slices, with 6 Tblsp. butter, ¼ cup lemon juice and salt and pepper to taste and sweat them over moderate heat, covered with a buttered round of wax paper and the lid, for 5-7 minutes, or until they are tender. In a heated large bowl, toss the broccoli with the mushrooms and the cooking liquid and add butter, salt and pepper to taste.

BROCCOLI IN CHEESE SAUCE

preparation time: 5-10 minutes

cooking time: 20 minutes

serves: 8

Linda Richards

Tao Mu

2 large or 3 small bunches
    broccoli
3 Tblsp. butter
½ cup flour
2½-3 cups milk (may use
    powdered milk, mix as
    directed)

1 cup grated cheddar cheese
Salt and pepper, to taste

Wash broccoli thoroughly and cut into about 1-inch pieces. Thick stems should be split to cook evenly. Steam in lightly salted water until crisp-tender. Don't overcook! While broccoli is cooking make cheese sauce: melt butter in medium saucepan or double boiler if available (take care not to brown butter) add flour and remove from heat, stirring with wire whip until smooth. Add ½ milk, whipping until smooth. Return to heat and cook gently until thickened, stirring constantly. Add cheese and enough milk to make a sauce of medium consistency (add more milk if necessary). Drain broccoli thoroughly and mix with enough sauce just to coat. Serve in a heated serving dish with remainder of sauce in a heated sauce boat for guests to add as they desire.

CREAMY BROCCOLI BAKE

preparation time: 25 minutes   *Liz Thomas*
cooking time: 20 minutes   *Raby Vaucluse*
serves: 6

2 (10 oz.) frozen broccoli (or   ½ cup shredded cheddar cheese
  cauliflower), cooked and   1 cup Bisquick
  drained   ¼ cup margarine
1 can cream of mushroom soup
¼ cup milk

Cook and drain broccoli. Beat soup and milk until smooth. Pour over broccoli in round casserole. Sprinkle with cheese. Mix baking mix and margarine until crumbly. Sprinkle over cheese. Bake at 400° for about 20 minutes or until browned.

CARROTS AMARETTO

preparation time: 10 minutes   *Gabrielle Thompson*
cooking time: 15 minutes   *Satori*
serves: 8

1 lb. carrots, cleaned, sliced ⅛"   2 oz. blanched almonds, sliced
  thick   ¼ cup amaretto
2 oz. butter   Salt and pepper

Saute carrots in butter in large skillet until lightly cooked. Add almonds. When carrots are softening, but still have crunch, add amaretto and warm through. Season with salt and pepper. So easy!!

STIR FRY CARROTS

preparation time: 15 minutes   *Harriet Beberman*
cooking time: 10 minutes   *Avatar*
serves: 4-6

6-8 carrots, coarsely grated   Salt and pepper, to taste
1 onion, coarsely grated   Pinch sugar
1 clove garlic   2 Tblsp. margarine

Coarsely grate carrots and onion. Stir fry in melted, sizzling margarine with garlic. Cook until tender in skillet, stirring with spatula as you would any stir fry dish. Add salt, pepper, and sugar. Stir once again and serve.

TARRAGON & CARROTS

preparation time: 10 minutes   *Betsi Dwyer*
cooking time: 20 minutes   *Malia*
serves: 6

10-12 carrots   **1 tsp. tarragon**
4 Tblsp. butter   **1 clove garlic, crushed**

Clean and slice carrots into ¼" slices. Steam just till al dente. Melt butter in a skillet. Add tarragon and garlic. Stir in carrots and quickly saute until carrots are well coated.

ORANGE CARROTS

preparation time: 10 minutes   *Gill Gase*
cooking time: 20 minutes   *M/V Polaris*
serves: 8

2 lbs. carrots   **2 Tblsp. butter**
4 Tblsp. orange juice concentrate

Peel carrots and slice into julienne strips. Boil in salted water till tender. Drain and toss in butter. Add 4 Tblsp. orange concentrate and stir over low flame till heated through.

QUICK FRIED CABBAGE

preparation time: 5 minutes   *Jan Robinson*
cooking time: 10 minutes   *Vanity*
serves: 4

½ medium white cabbage   **Oil for deep frying**
½ cup milk   **Salt, optional**
½ cup flour

Cut off damaged outer leaves, wash cabbage. Cut out hard central stalk and finely shred cabbage. Dip a few cabbage shreds at a time in the milk, then toss them in the flour on a sheet of waxed paper. Heat the oil to 375° in a deep fryer; put a few shreds of coated cabbage into basket and deep-fry for 1-2 minutes until crisp and golden. Drain on paper towels. Keep warm. Continue process until all cabbage is cooked. Sprinkle with salt and serve at once. Excellent with broiled fish or meat.

FLUFF'S CAULIFLOWER

preparation time: 15 minutes   Jan Robinson
cooking time: 20 minutes   *Vanity*
serves: 4-6

1 cauliflower   2 Tblsp. chopped parsley
2 hard-cooked eggs   Salt and black pepper
4 Tblsp. butter   Lemon juice
¾ cup dry bread crumbs

Remove outer leaves and thick stalk base of cauliflower and wash thoroughly. Steam cauliflower for 10-15 minutes or until just tender. Drain and cover to keep warm. Peel eggs separating yolks from whites. Put yolks through sieve and chop whites. Melt butter and saute breadcrumbs until crisp. Remove from heat. Add parsley, salt, pepper and lemon juice to taste. Place cauliflower in a warm serving dish, sprinkle with bread crumbs and garnish with yolks on top surrounded by chopped whites in a flower pattern.

CORN AND OYSTER CASSEROLE

preparation time: 5 minutes   Debbie Olsen
cooking time: 30-40 minutes   *Kingsport*
serves: 8

1 can corn   1 pint fresh oysters, drained and
1 cup cracker crumbs   minced
1 cup sour cream

Mix all together and bake in oblong baking dish at 375° until golden brown. This great recipe comes from Roy's mother, Mary Love Olsen.

GRILLED RED ONIONS

preparation time: 10 minutes   Jean Thayer
marinating time: 1 hour   *Paradise*
cooking time: 10 minutes
serves: 6

6 medium red onions   2 tsp. dried thyme
¾ cup vegetable oil

Peel and slice onions in half lengthwise. Trim top of onion. In a shallow dish, combine oil and thyme. Marinate onions in the oil for at least 1 hour, turning them once or twice to coat. Grill for ten minutes over medium coals and serve with Fool Proof Barbecued Steaks.

CORN PUDDING

preparation time: 15 minutes   *Ann Glenn*
cooking time: 1 hour   *Encore*
serves: 6-8

½ cup Wondra flour
½ cup liquid Parkay
2 Tblsp. sugar
1 tsp. salt
Pepper, to taste

2 cups milk
1 can cream style corn
1 can whole kernel corn, drained
3 eggs, separated

Over low heat, combine flour and Parkay until smooth. Add sugar, salt and pepper, then add milk slowly to make white sauce. Stir in both cans of corn and egg yolks. Beat egg whites until stiff, then fold into mixture. Pour into shallow pan, 9 x 13, and bake at 350° for an hour. Remove from oven and let set 10 minutes before serving.

EGGPLANT AND TOMATO RAGOUT

preparation time: 40 minutes   *Margaret Benjamin*
cooking time: 35 minutes   *Illusion II*
serves: 6

2 large or 3 small eggplants
1 large onion
2 oz. butter
salt and black pepper

1 tsp. castor sugar
16 oz. can tomatoes
½ tsp. basil

Peel eggplant into ½-inch slices. Sprinkle with salt and let stand for 30 minutes. Rinse with cold water and pat dry. Slice onion and saute in butter for 5 minutes, add eggplant and saute for a few more minutes. Stir in tomatoes and juice and seasonings. Cover and simmer for 30 minutes.

JEANO'S ONION CASSEROLE

preparation time: 10 minutes   *Jan Robinson*
cooking time: 20-25 minutes   *Vanity*
serves: 6-8

6-7 Spanish onions
2 cans cream of mushroom soup
Butter

Parmesan cheese
Salt and pepper, to taste

Thickly slice the onions and place in a Pyrex dish 12 x 7 x 1½. Pour the 2 cans of soup, undiluted, over the onions and sprinkle with salt, pepper, parmesan cheese and generous pats of butter. Cover and bake at 350° for 20-25 minutes. Serve hot or luke warm. As Jeano says, "There is never any left over!"

SAUTÉED RED ONIONS AND CARROTS

preparation time: 15 minutes   Nancy Thorne
cooking time: 20 minutes   Gullviva
serves: 6

2 Tblsp. butter   **3 red onions, thinly sliced**
2 lbs. young carrots   **Pinch basil**

Melt butter. Add carrots, diagonally sliced ¼ inch thick, and onions. Cover and cook over low heat 15 to 25 minutes. Add basil, cover pan and cook till fork tender.

CREAMED ONIONS

preparation time: 10 minutes   Terry Boudreau
cooking time: 40 minutes   High Barbaree
serves: 8

2½ lbs. small white onions   **2 cups Mornay sauce (see index)**
½ bay leaf   **Grated nutmeg**

In a saucepan combine 2½ lbs. small white onions, peeled, with ½ bay leaf, and enough salted cold water to cover them by 2 inches; bring to a boil over moderate heat, and simmer the onions for 25 minutes, or until they are tender. Drain the onions, in another saucepan combine them with 2 cups Mornay sauce and freshly grated nutmeg to taste; heat, stirring, until it is heated through. Transfer the onion mixture to a heated platter.

PETIT POIS PEAS AND SCALLIONS

preparation time: 15 minutes   Cheryl Anne Fowler
cooking time: 15 minutes   Vanity
serves: 6

2 (10 oz.) pkgs. frozen petit pois   **2 Tblsp. butter**
    peas   **4 Tblsp. flour**
10-12 scallions   **Salt and black pepper**
1¼ cups chicken stock

Wash, cut stems off and trim outer leaves of scallions leaving about 1 inch of green on each bulb. Combine scallions and stock in pan and bring to simmering point. Gently cook until softened and add peas, cooking till tender over low heat. Combine butter and flour and make a paste. When peas are just cooked, gradually crumble the paste into the mixture and stir carefully until stock thickens. Season to taste with salt and pepper and serve immediately.

LETTUCE AND GREEN PEAS

preparation time: 15 minutes   Jan Robinson
cooking time: 15-20 minutes   Vanity
serves: 6-8

10 outer Romaine lettuce leaves   1 Tblsp. sugar
2 (10 oz.) pkgs. small frozen peas   ½ cup water
1 large onion, chopped fine   3 sprigs parsley
4 Tblsp. butter   1 slice bacon, cooked, crumbled

Place lettuce, peas, onion, 2 Tblsp. butter, sugar, parsley and water to cover in a heavy saucepan. Mix gently and cover. Simmer until peas are tender, about 15 minutes. Remove from heat; remove parsley. Add remaining butter and return to heat shaking pan until butter is melted and almost all liquid has evaporated. *To serve, sprinkle with crumbled bacon.*

PERO'S FAVORITE PLANTAINS

preparation time: 20 minutes   Cheryl Anne Fowler
cooking time: 12-15 minutes   Vanity
makes: 12-18

3 green plantains   2 Tblsp. salt
4 cups water   Oil for deep frying
2 cloves garlic, peeled and
   crushed

Peel and cut plantains into diagonal slices 1 inch thick. Add garlic and salt to water and soak plantain slices for 15 minutes. Drain well and deep-fry in fat at 350° about 7 minutes. Remove from pan and place on paper towel. Fold the towel over slices and pound flat with palm of hand. Dip in the salted water again and remove immediately. Drain thoroughly on absorbent towel. Again deep-fry at 375° until crisp and golden. Remove, drain and lightly salt.

POTATOES ST. JOHN

preparation time: 30 minutes   Frances Bryson
cooking time: 15-20 minutes   Amoeba

Mashed potatoes   Cornflake crumbs
Cheddar cheese (or your favorite)   Butter, melted

Prepare mashed potatoes in advance and let cool. Form in balls adding a piece of cheddar cheese (or cheese of your choice) to the center of the ball. Roll balls first in cornflake crumbs, then in melted butter. Place in baking pan so balls do not touch and bake for 15-20 minutes at 300°. This dish can be prepared in advance and refrigerated or frozen.

POTATOES AU GRATIN IN FOIL

preparation time: 25 minutes   *Betsi Dwyer*
cooking time: 40 minutes   *Malia*
serves: 8

8 baking potatoes, skinned   **Butter**
2 small onions   **Tin foil**
½ lb. grated cheese   **Salt**
¼ cup milk   **Pepper**

Slice skinned potatoes into ¼" pieces. Place pats of butter onto a foil
sheet, add onion slices, cheese, potatoes, salt, pepper. Repeat until 3 layers are complete. Add ¼ cup milk and wrap twice in tin foil. Cook over
hot coals for 40 minutes, 20 on each side.

GARLIC POTATOES

preparation time: 20 minutes   *Maureen Stone*
cooking time: 30-40 minutes   *Cantamar IV*
serves: 6-8

6 medium potatoes   **1 cup whipping cream or**
16 oz. cheddar cheese, grated   **evaporated milk**
                                   4-6 garlic cloves, crushed

Peel and boil potatoes until tender, drain and cool. Slice potatoes and arrange in greased baking dish. Sprinkle with salt and pepper. Top with ⅓
of cheese. Pour ½ of cream over cheese. Again top with cheese, pour remaining cream and top with remaining cheese. Sprinkle garlic evenly over
the top. Bake until top is browned and crisp, 30-40 minutes at 350°.

HASSELBACK POTATOES

preparation time: 20 minutes   *Gill Case*
cooking time: 45 minutes   *M/V Polaris*
serves: 8

8 medium baking potatoes   **Butter**
Parmesan cheese, grated   **3 Tblsp. dried breadcrumbs**

Scrub potatoes and slice downwards at ¼" intervals without cutting all
the way through. Put in a pan in cold, salted water, bring to a boil and
cook 4-5 minutes. Meanwhile, preheat oven to 450°. Butter an ovenproof
dish. Drain potatoes, place sliced side upwards in dish. Dot with butter
and bake 30 minutes, basting occasionally. Mix parmesan with crumbs,
scatter over potatoes and bake 10-15 minutes more.

POMMES DE TERRE ANNA

preparation time: 10 minutes   Shirley Fortune
cooking time: 50 minutes   Maranatha
serves: 6

6 medium baking potatoes, thinly   Pepper
**   sliced   ⅓ cup melted butter**
1 tsp. salt

Arrange layer of potato slices in bottom of generously greased 9-inch pie
pan. Sprinkle with a small amount of salt and pepper. Repeat until all potatoes are layered, sprinkling each layer with salt and pepper. Pour butter
over all and bake in 400° oven until tender, about 50 minutes. Loosen
edges and bottom of potatoes with a wide spatula (this is important).
Place inverted dish over plate; invert potatoes onto plate. Garnish with
parsley sprigs in the center. Cut in wedges to serve.

GETTING FANCY POTATOES

preparation time: 20 minutes   Kathleen McNally
cooking time: 20 minutes   Orka
serves: 6

¼ cup mayonnaise (or with half   2 slices bacon, crisp and crumbled
**   sour cream)   2 Tblsp. chopped onion**
2 Tblsp. dijon mustard   2 hard-boiled eggs, chopped
2 Tblsp. white wine   1 clove garlic, minced
6 large fresh mushrooms   1-2 large fresh tomatoes
¼ cup favorite cheese, grated
6 medium potatoes, boiled

Mix mayonnaise, white wine and mustard in small saucepan over low
heat. Simmer 5 minutes, keep warm. Make mashed potatoes with 6
spuds adding, when mashed, the cheese, bacon bits, chopped onion, garlic and egg (some people prefer to have already cooked the onion until
soft before adding). Separate carefully the mushroom stems from caps.
Set caps aside intact. Chop stems, add to potato mixture. Now slice the
tomatoes into ¼-inch slices. On plate place tomato slice, mushroom cap
(hollow up) followed by large spoonful of potatoes. Cover with sauce and
a dash of paprika. You can skip tomato slice and mushroom cups and
just serve as Baked Potato Casserole Gratin. Good enough for Royalty!

PARMESAN POTATOES

preparation time: 10-15 minutes   Ann Glenn
cooking time: 45 minutes   Encore
serves: 6-8

6-8 large, unpeeled potatoes, well   **1 tsp. salt**
   scrubbed and cut into wedges   **¼ tsp. garlic powder and paprika**
½ cup oil   **¼ tsp. ground pepper**
2 Tblsp. parmesan cheese, grated

Preheat oven to 375°. Arrange potato wedges in 9 x 13 baking pan,
making first layer skin-side down, and a second layer of wedges in between the first. Add cheese and seasonings to the oil and baste potatoes
liberally. Optional: Sprinkle with additional cheese. Bake for 45 minutes.
Note: These potatoes may be "finished" in the microwave, or cooked en-
tirely by that method, in which case 15-20 minutes is a *guesstimate* of total cooking time.

BAKED POTATO CHIPS

preparation time: 15 minutes   Kathleen McNally
cooking time: 20 minutes   Orka
serves: 6

3½ lbs. red or white boiling   **6 Tblsp. (¾ stick) butter, melted**
   potatoes, cut crosswise into   **Salt**
   ⅛-inch thick slices   **Freshly ground pepper**

Position racks in upper and lower third of oven and preheat to 500°.
Lightly grease 2 baking sheets. Arrange potato slices in single layer on
prepared baking sheets. Brush generously with butter. Bake 7 minutes.
Switch pan positions and continue baking until potatoes are crisp and
browned around edges, about 7-9 minutes. Transfer to heated platter.
Sprinkle with salt and freshly ground pepper. *Serve immediately.*

• *Cook cauliflower in half milk an half water to keep white and elimi-
nate cooking odor.*

POTATO LYONAISE

preparation time: 15 minutes
cooking time: 30 minutes
serves: 12

Paula Taylor
Viking Maiden

8 potatoes, peeled and halved
6 large onions, sliced
4 oz. butter

2 cloves garlic, crushed
1 Tblsp. dried parsley

Boil potatoes until cooked but firm, drain potatoes, melt butter in pan, add garlic, onions and parsley to pan with potatoes. Cook, turning frequently until potatoes are brown and onions are cooked. *Serve garnished with sprigs of parsley.*

GOLDEN CURRIED POTATOES

preparation time: 30 minutes
cooking time: 10 minutes
serves: 6

Terry Boudreau
High Barbaree

½ cup finely chopped onion
¼ cup butter
2 tsp. curry powder
3 cups cooked potatoes, cut in
    thin strips

½ cup chicken bouillon
1 tsp. salt
Dash pepper
1½ tsp. lemon juice

Saute onion in butter until transparent. Stir in curry powder and cook about 1 minute. Add potatoes and stir gently until butter is absorbed. Pour bouillon over potatoes and simmer until liquid is absorbed (about 5 minutes). Add seasonings and cook 1 minute longer.

BAKED ACORN SQUASH SLICES

preparation time: 10 minutes
cooking time: 50 minutes
serves: 4

Launa Cable
Antiquity

1 acorn squash, about 2 lbs.
½ cup orange juice
¼ cup firmly packed light brown
    sugar

3 Tblsp. melted butter
1 tsp. salt

Cut squash into ¾-inch thick crosswise slices, remove seeds. Arrange slices in shallow baking dish. Combine orange juice, sugar, butter and salt in a cup, blend well. Pour over squash slices. Bake in 400° oven for 50 minutes.

SCRAMBLED TOFU & POTATOES

preparation time: 10 minutes   *Ginger Outlaw-Fleming*
cooking time: 20 minutes   *Coyaba*
serves: 4

1 pkg. 10 oz. tofu   **Salt**
1 onion   **Pepper**
1 small carrot   **Tumeric or curry powder**
1 potato   **Soy sauce**

Dice onion, quarter carrot and cube potato. Saute in small amount of water or oil till soft, about 10 minutes. Crumble tofu in hands till it's in small pieces and add to sauteed mixture. Cook on low heat for 5-10 minutes, add tumeric (or curry for an exotic taste) at the end of cooking. It adds the color of make-believe scrambled eggs and if you didn't know it was tofu, you would actually swear it was eggs. Add salt and pepper to taste.

PORGIE'S SQUASH

preparation time: 20 minutes   *Jan Robinson*
cooking time: 40-45 minutes   *Vanity*
serves: 4-6

3 cups grated squash, yellow   **½ tsp. oregano**
   crookneck or zucchini   **1 cup Bisquick**
1 small onion, grated or chopped **4 eggs — lightly beaten**
½ cup oil   **4 oz. parmesan cheese**

Grease small 8 x 8 casserole. Blend all ingredients. Pour into casserole dish and bake at 325° for 40-45 minutes. Should there be any leftover, slice in finger sized pieces and freeze. Makes a great quick hors d'oeuvre — pop in hot oven, heat quickly and serve. Also good served as a light vegetable lunch accompanied by a crisp green salad and rolls.

SCHOONER SPINACH

preparation time: 5 minutes   *Gay Thompson*
cooking time: 15 minutes   *Satori*
serves: 8

2 (9 oz.) pkgs. chopped, frozen   **1¼ cup sour cream**
   spinach
1 pkg. Lipton dry onion soup mix

Cook spinach according to package directions. 60 seconds before spinach is finished cooking, add dry onion soup mix. Drain. Add sour cream and heat through. *Serve immediately.*

DELANY'S CREAMED SPINACH

preparation time: 10 minutes   Louella Holston
cooking time: 20 minutes   Eudroma
serves: 6-8

2 (10 oz.) pkgs. frozen spinach,   2 cups milk
   cooked and drained well   ½ cup grated parmesan cheese
1 onion, minced   Dash of nutmeg
2 Tblsp. butter   Salt and pepper (white)
4 Tblsp. flour

Saute onion in butter. Add flour and stir until all butter is absorbed. Add
milk slowly and whisk to keep smooth. Cook over medium heat until
thick. Add cheese and seasonings. Stir cheese mixture into spinach and
heat thoroughly, being careful not to burn.

SPINACH/ARTICHOKE

preparation time: 10-15 minutes   Ann Glenn
cooking time: 25 minutes (or micro about 10 minutes)   Encore
serves: 8

2 pkgs. (10 oz.) chopped spinach   1 Tblsp. lemon juice
1 can artichokes (hearts or   Cracker crumbs
   bottoms)   Butter
½ cup butter or liquid Parkay
1 pkg. (8 oz.) cream cheese,
   softened

Preheat oven to 350°. Grease casserole. Cook spinach and artichoke
briefly; drain well. Chop artichoke and place in bottom of casserole. Combine spinach with cream cheese and lemon juice, blending well. Spread
spinach mixture over artichoke layer. Top with cracker crumbs, dot with
butter and bake at 350° for about 25 minutes (or microwave about 10
minutes).

SPINACH & ARTICHOKE CASSEROLE

preparation time: 15 minutes
cooking time: 30 minutes
serves: 12-14

Pamela McMichael
Pieces of Eight

**5 (10 oz.) pkgs. frozen chopped
    spinach**
1 (14 oz.) can artichoke hearts
2 cups sour cream

1 envelope onion soup mix
Salt, to taste
Pepper, to taste

Cook spinach and drain well. Quarter artichokes. Combine all ingredients. Bake in buttered 2 qt. casserole at 350° for 30 minutes.

Quick, easy and elegant!

MUSHROOMS FLORENTINE

preparation time: 15 minutes
Cooking time: 20 minutes
serves: 6-8

Maureen Stone
Cantamar IV

1 lb. mushrooms
**2 (10 oz.) frozen spinach or 1 lb.
    fresh, cooked and drained**
1 tsp. salt

¼ cup chopped onion
¼ cup melted butter
1 cup grated cheddar cheese
Garlic salt

Clean mushrooms. Slice off stems and saute with caps until brown. Line a shallow 10-inch casserole with butter. Sprinkle with ½ cup cheese and top with all the mushrooms. Sprinkle with garlic salt and top with remaining cheese. Bake at 350° for 20 minutes until cheese is melted.

DEVILED TOMATOES

preparation time: 5 minutes
cooking time: 5 minutes
serves: 8

Maureen Stone
Cantamar IV

4 large tomatoes
Salt
½ cup softened butter
Worcestershire sauce
2 tsp. grated onion

1 tsp. dry mustard
½ tsp. tabasco sauce
**2 Tblsp. chopped fresh parsley or
    1 Tblsp. dried parsley**

Cut tomatoes in half and place cut side up in shallow baking dish. Sprinkle lightly with salt. In a small bowl, whisk remaining ingredients together. Spoon over each tomato. Place under broiler for about 6-8 minutes.

FESTIVE BAKED TOMATOES

preparation time: 5 minutes
cooking time: 15 minutes
serves: 4

D. J. Parker
Trekker

4 large ripe tomatoes
1 Tblsp. mustard

4 Tblsp. minced green sweet
peppers
4 Tblsp. minced onion

Rinse tomatoes. Cut out stem area and cut in quarters without severing bottom. Carefully spread sides of each quarter with mustard. Place one Tblsp. each green pepper and onion inside each tomato. Place in shallow baking dish and bake in 350° oven for 15 minutes. *Serve hot. Excellent with rice.*

TOMATOES PROVENCALE

preparation time: 10 minutes
cooking time: 25 minutes
serves: 6

Betsi Dwyer
Malia

6 large tomatoes
⅓ cup bread crumbs
5 Tblsp. grated parmesan
2 Tblsp. minced parsley

1 clove garlic, minced
Olive oil
Salt, to taste
Pepper, to taste

Halve tomatoes and trim ends so they stand level. Combine all other ingredients and divide among tomatoes. Drizzle with oil. Bake at 375° for 20-25 minutes.

TOMATOES FLORENTINE

preparation time: 10 minutes
cooking time: 15 minutes
serves: 6

Chris King
Sunrise

6 small tomatoes
1 pkg. frozen creamed spinach

Parmesan cheese

Preheat oven to 375° or microwave 5 minutes. Cut tops off the tomatoes. Hollow out the tomato shells removing pulp and seeds. Fill tomato shells with creamed spinach and sprinkle with parmesan cheese. Heat in oven approximately 15 minutes, checking for doneness.

MUSHROOM STUFFED TOMATOES

preparation time: 20 minutes
cooking time: 10 minutes
serves: 6

Shirley Fortune
Maranatha

6 whole medium tomatoes
½ cup sliced fresh mushrooms
3 Tblsp. Butter
2 Tblsp. tomato paste or ketchup
2 Tblsp. water
½ tsp. salt

⅛ tsp. pepper
1 egg yolk, beaten
¼ cup parmesan cheese
Butter for topping
¼ cup bread crumbs

Cut stem ends off tomatoes and scoop out seeds, leaving a firm wall. Filling: Saute mushrooms 3 minutes. In another pan combine tomato paste, water, salt and pepper and cook until well blended and let cool approximately 4 minutes. Add egg yolk gradually a spoonful at a time. Add tomato-egg mixture to mushrooms and stuff tomatoes. Sprinkle tops with crumbs and cheese top with a dot of butter. Bake at 425° for 10 minutes — do not overcook as tomatoes become mushy. An attractive vegetable garnish with roasts.

ZUCCHINI SQUARES

preparation time: 15 minutes
cooking time: 20-25 minutes
serves: 8

Ann Glenn
Encore

1 cup Bisquick
½ cup grated Parmesan cheese
2 Tblsp. snipped parsley
½ tsp. salt & ½ tsp. butter salt
½ tsp. oregano leaves
Fresh ground pepper, to taste

3-4 small zucchinis, thinly sliced
1 medium onion, chopped (red onion preferably)
1 large clove garlic, pressed
½ cup vegetable oil
4 eggs

Combine the dry ingredients in a large bowl: Bisquick, cheese, parsley, oregano, salt & butter salt, and pepper. In smaller bowl, stir together 4 eggs and vegetable oil. Mix wet and dry ingredients, then add zucchinis, onion and garlic. Spread in greased 9 x 13 pan. Bake until lightly browned, about 25 minutes at 350°.

This recipe is particularly versatile. In addition to being a side dish for lunch or dinner, if cut diagonally into small diamond shapes, it will make several dozen delicious hot hors d'oeuvres.

GRATED ZUCCHINI

preparation time: 20 minutes   *Betsi Dwyer*
cooking time: 10 minutes   *Malia*
serves: 6

2 medium zucchini squash   **Salt and pepper, to taste**
2 medium yellow squash   **4 Tblsp. butter**
1 tsp. garlic powder

Grate squash. Sprinkle with salt and set aside for 10 minutes. Squeeze excess water from squash. Melt butter over low heat. Add spices and then squash and cook over medium high heat for 5 minutes.

PARMESAN ZUCCHINI FLARES

preparation time: 10 minutes   *D. J. Parker*
cooking time: 15 minutes   *Trekker*
serves: 4

4 fresh small zucchini   **Grated fresh parmesan**
2 Tblsp. butter, melted   **Water**

Wash and trim zucchini. Cut each in half crosswise. Cut each half in thirds lengthwise almost to end. Be careful not to sever the ends so that each piece somewhat resembles a fan. Place in water and bring to boil. Cook just until tender (do not overcook). Drain. Using a fork, flare thirds to fanlike. Pour melted butter over zucchini and sprinkle with cheese. *Serve hot.*

SPINACH FETTUCCINE

preparation time: 10 minutes   *Gay Thompson*
cooking time: 15-20 minutes   *Satori*
serves: 8

4-6 ozs. thin spaghetti or   **½ Tblsp. coarse ground black**
  fettuccine noodles   **pepper**
1 (9 oz.) pkg. chopped frozen   **½ cup grated parmesan cheese**
  spinach, thawed, drained   **4-6 ozs. sour cream**

Cook noodles according to package directions. When they are ¾ done (still hard in center), add drained, thawed spinach (squeeze out excessive water). When noodles are done, drain off water well and add sour cream, pepper and parmesan cheese and blend quickly. *Serve immediately.*

ZUCCHINI PESTO

preparation time: 10 minutes   *Kathleen Leddy*
cooking time: 6 minutes   *Carriba*
serves: 6

4 zucchini, medium sized,   **2 Tblsp. parmesan, grated**
   julienned   **2 cloves garlic, finely chopped**
⅓ cup basil leaves   **Dash salt and fresh ground**
¼ cup olive oil   **pepper, to taste**
2 Tblsp. walnuts

Combine basil, oil, nuts, garlic, and cheese in skillet and heat for one minute after oil is hot. Do not overcook. Remove from pan and add 1 Tblsp.
oil to medium-high heat, add zucchini and cook 3 minutes, tossing frequently. Season with salt and pepper. Stir in pesto, warm one minute and
serve.

SUMMER SQUASH CASSEROLE (Zucchini)

preparation time: 20 minutes   *Terry Boudreau*
cooking time: 30 minutes   *High Barbaree*
serves: 6

3 cups zucchini   **1 beaten egg yolk**
¼ cup cultured sour cream   **1 Tblsp. chopped chives**
1 Tblsp. butter   **Au gratin topping**
1 Tblsp. grated cheese   **(Dry bread crumbs, dots of butter**
½ tsp. salt   **and grated cheese)**
⅛ tsp. paprika

Cut 3 cups zucchini into small pieces. Simmer the squash, covered, until
tender, for about 6-8 minutes, in a small amount of boiling water. Shake
the pan to keep from sticking. Drain well. Combine sour cream, butter,
grated cheese, salt, and paprika in another saucepan, stir until the cheese
is melted. Remove from heat and stir in 1 beaten egg yolk, and the
chopped chives. Cover the zucchini with Au gratin and brown it in a
heated oven at 375°.

MANICOTTI

preparation time: 30 minutes
cooking time: 20 minutes
serves: 6

Chris Kling
Sunrise

1 box manicotti pasta
1 lb. ricotta cheese
8 oz. grated mozzarella cheese
8 oz. cream cheese, softened
3 Tblsp. minced parsley

2 egg yolks
½ tsp. grated nutmeg
1 jar spaghetti sauce
Parmesan cheese

Cook manicotti according to directions on box. Drain and cover with cold water. In a mixing bowl combine cheeses, parsley, egg yolks, and nutmeg. Spread ½ spaghetti sauce in the bottom of a large baking dish. Stuff manicottis with cheese filling and place in baking dish. Pour remaining sauce over and sprinkle with parmesan cheese. Bake in 375° oven for 20 minutes.

*FETTUCCINE GENE

preparation time: 10 minutes
cooking time: 20 minutes
serves: 6

Lori Moreau
Alberta Rose

1 pkg. linguine noodles
4 oz. butter, melted
2 egg yolks

⅔ cup light cream
6 oz. grated Swiss
Salt

Leave egg yolks, cream and cheese out until room temperature. Cook noodles and drain well. In large pot over low heat, toss noodles with ingredients in the order listed (one item at a time). When cheese is melted and tossed thoroughly, salt to taste and serve immediately.

*I call this Fettuccine Gene because he does all the tossing; I'm too short to reach over the top of the pot!!

NOODLES ALFREDO

preparation time: 10 minutes
cooking time: 15 minutes
serves: 6-10

Harriet Beberman
Avatar

1 lb. medium egg noodles
4 to 8 quarts water
⅛ cup salt
½ lb. sweet butter, softened

2 cups grated parmesan cheese
½ cup heavy cream (room
    temperature)
Fresh ground pepper

Gradually add noodles and salt to rapidly boiling water so that water continues to boil. Cook uncovered, stirring occasionally until tender. Drain in colander. Place butter in hot 4 quart casserole dish. Add noodles and toss gently. Add cheese and toss again. Pour in cream, toss. Sprinkle with freshly ground pepper.

FETTUCCINE ALLA ROMANO

preparation time: 20 minutes
cooking time: 25 minutes
serves: 6-8

Betsi Dwyer
Malia

1½ cups heavy cream
1½ cups freshly grated parmesan
    (5½ oz.)
½ cup butter
2 egg yolks, beaten

1 lb. fettuccine
1 cup cooked peas or bite-sized
    pieces of asparagus
⅔ cup finely shredded prosciutto

Heat cream over low heat. When it begins to simmer, stir in cheese, bit by bit. Continue stirring for 10 minutes. Add butter bit by bit. When butter is incorporated and sauce smooth, remove from heat and beat small amount into yolks. Return yolk mixture to sauce; stir with whisk. Add peas and prosciutto to cooked pasta. Pour sauce over all and toss till well coated.

FETTUCCINE VENETIAN STYLE

preparation time: 25 minutes
cooking time: 18 minutes
serves: 8

Maureen Stone
Cantamar IV

1½ medium onions, sliced
2 Tblsp. sweet butter
1 cup dry white wine
1 tsp. salt
1½ lbs. sole fillets or other mild fish
½ lb. shrimp, shelled, deveined, cut into small pieces
1 cup light cream

½ tsp. white pepper
Pinch curry powder (optional)
1 lb. fettuccine
3 Tblsp. salt
2 Tblsp. freshly grated Parmesan cheese
2 Tblsp. freshly grated Romano

In large skillet, saute onions in one tablespoon of butter till soft. Add wine and ½ tsp. of salt. Stir in about one pound of the fish cubes and half the shrimp. Cook for 8-10 minutes. Put the rest of the raw fish and shrimp in blender with 2 tablespoons of the broth from the cooked onion-fish mixture. Blend to a paste, remove from blender and fold in cream, pepper, curry powder. Simmer this mixture over a low flame, adding remaining salt. Cook fettuccine al dente in salted boiling water. Drain. Butter a large baking dish. Spread a layer of fettuccine on the bottom, then a layer of the cooked fish cubes and shrimp then a layer of the paste mixture. Repeat till all ingredients are used up, ending with fettuccine. Combine cheeses, sprinkle on top and broil until the top is browned, about 8 minutes.

LINGUINI & WHITE CLAM SAUCE

preparation time: 15 minutes
cooking time: 20 minutes
serves: 6

Trish Penn
Daddy Warbucks

2 (16 oz.) cans clams with juice or 2 dozen cooked clams with juice
3-4 cloves garlic
2 Tblsp. olive oil

4 Tblsp. parsley, chopped
½ cup milk
1 box linguini
Water and cornstarch or flour, to thicken

Heat oil in saucepan. Add chopped garlic. Cook until golden. Add clams in liquid and parsley. Cook for 10 minutes. Add milk. Cook 5 minutes. Just before serving add thickening (water and cornstarch or flour) just enough to thicken to a gravy state. *Serve over cooked linguini.*

TREKKER'S FETTUCCINE MARINARA

preparation time: 30 minutes   D. J. Parker
cooking time: 20 minutes   Trekker
serves: 4

1 egg yolk
1 cup sour cream
¼ lb. butter
1 large conch, cleaned,
    tenderized, minced

6 green onions (spring)
1 cup lobster pieces, cooked
8 oz. spinach egg noodles
⅓ cup grated parmesan cheese

Blend egg yolk and sour cream. Set aside to become room temperature. Melt butter in small saucepan. Mince green onions including part of tops. Add onions and conch to butter and saute until onions become transparent; add lobster, stirring lightly until lobster is heated. Cook spinach noodles according to directions on container. Drain noodles thoroughly. Toss butter and seafood mixture with noodles. Gently add sour cream and parmesan cheese to noodles. *Serve immediately with sprinkle of parmesan cheese on top. Garnish with tomato and cucumber slices.*

COUSCOUS

preparation time: 5 minutes   Ann Glenn
cooking time: 20 minutes   Encore
serves: 6

1 medium onion, chopped
2 Tblsp. butter
2 cups liquid (1 can chicken broth
    and liquid from garbanzos)
1 cup couscous

½ tsp. salt and ground cumin
1 can garbanzos, drained
1 cup or more, cubed ham or
    chicken (optional)

In medium saucepan, saute onion in butter. Add liquid and garbanzos and bring to the boil. Stir in couscous, salt and cumin. Return to boil for 1-2 minutes. Optional: stir in cubed meat. Remove from fire and let set, covered, 10 minutes. Fluff with fork before serving.

It is fun to introduce guests to what will be, for many of them, a new and unusual taste. Couscous is an exotic substitute for rice, and the cumin gives it a flavor all its own.

• *Rice will remain white when cooking by adding one teaspoon lemon juice or one tablespoon vinegar to water.*

BULGAR PILAF

preparation time: 15 minutes
cooking time: 25 minutes
serves: 4

Pamela McMichael
Pieces of Eight

½ cup medium bulgar
2 medium carrots
1 medium onion

2 tsp. vegetable oil
1 tsp. salt
1 cup chicken broth

Wash and drain bulgar wheat. Pare carrots and peel onion. Dice both. Heat oil and add vegetables, saute till softened. Add bulgar and coat well with oil. Add stock and heat till boiling. Simmer covered until bulgar is tender and liquid has been absorbed — 25 minutes. *Serve.*

GARLIC BROWN RICE

preparation time: 5 minutes
cooking time: 50 minutes
serves: 4

Maureen Stone
Cantamar IV

1 cup brown rice
1½ Tblsp. olive oil
3 garlic cloves, chopped fine

½ medium onion
1 tsp. salt
2 cups water

Fry 1 cup brown rice in 1½ Tblsp. olive oil and 2 garlic cloves, for a few minutes. Add 2 cups water and 1 tsp. salt. Bring to a boil. Place ½ of the cut onion on top and simmer, covered for 50 minutes.

WILD RICE FOR POULTRY

preparation time: 10 minutes
cooking time: 45-60 minutes
serves: 6

Terry Boudreau
High Barbaree

1 medium onion, chopped
2 Tblsp. butter
1 cup wild rice
4 cups chicken broth

¼ tsp. each: sage, thyme, and
   majoram
Salt, if needed

Saute the onion in butter in a heavy 3 qt. saucepan. Add the wild rice, chicken broth and spices. Bring to boil. Cover, reduce heat and simmer 45-60 minutes or until most of the liquid has been absorbed. Fluff rice with a fork. Salt if needed. Cook uncovered to evaporate any excess moisture. Makes about 4 cups rice.

NUTTY WILD RICE

preparation time: 20 minutes   *Nancy Thorne*
cooking time: 20 minutes   *Gullviva*
serves: 6

2 cups Uncle Ben's Long Grain &   ½ tsp. marjoram
  Wild Rice   1 tsp. salt
1 onion, chopped   1 cup pecans, chopped
1 cup celery, minced   ½ cup parsley, chopped
4 cups beef broth
1 stick butter (½ cup)

Cook rice in beef broth and let stand. Saute onion and celery in butter.
Add marjoram, rice and salt. Stir in chopped pecans and parsley.

RICE RISOTTO

preparation time: 5 minutes   *Adriane Biggs*
cooking time: 30 minutes   *Taza Grande*
serves: 8

1½ cups rice (uncooked)   1 tsp. salt
1 medium onion, minced   ¼ tsp. pepper
3 cups boiling chicken broth   1 cup grated parmesan cheese
1 tsp. parsley   2 oz. butter

Melt butter in medium-sized saucepan. Add onion and saute until tender.
Add boiling chicken broth, parsley, salt, pepper and rice. Cover and simmer for 20 minutes. Remove from heat and stir in cheese.

STIR-FRIED RICE

preparation time: 10 minutes   *Gill Case*
cooking time: 20 minutes   *M/V Polaris*
serves: 8

4 cups rice   1 large onion, diced
2 large eggs, beaten   1 green pepper, diced
Olive oil   2 Tblsp. soy sauce
4 oz. diced, cooked ham

Cook rice, drain well. Meanwhile stir-fry onion, pepper and ham in a large
skillet. Add rice and stir well, adding more oil if necessary. Add soy sauce
and mix well into rice. Pour eggs into rice in a thin stream, stirring all the
time. Continue stirring till all ingredients are hot and the eggs are set.
Rice can be set aside and reheated just before serving.

MEDITERRANEAN VEGETABLES & RICE

preparation time: 2 hours 20 minutes   *Ann Doubilet*
cooking time: 40 minutes   *Spice of Life*

2½ cups unpeeled eggplant, cut in
  ½ inch cubes — salted
8 Tblsp. olive oil
2-3 zucchinis, cut in half rounds
1 onion, sliced
1 green pepper, chopped
2 cloves garlic, chopped

1 cup uncooked rice
2½ cups chicken broth
Ground red pepper
2 tomatoes
2 Tblsp. lemon juice and fresh
  parsley

If time, cut eggplant into cubes, salt cubes and let dry on paper towel for 2 hours to dry out. Pat dry and saute in olive oil in batches over medium-high heat until golden brown. Remove from pan and drain. Add more oil and add all vegetables except tomato. Cook until soft. Add tomato and ground red pepper to taste. Cook until partly dry. Stir in rice, broth, lemon juice, salt and peppers. Cook until rice is done — about 25 minutes. Stir in eggplant. Add minced parsley. You can add curry powder at time of adding tomatoes if you prefer. Recipe can be made without drying eggplant, but it will not brown as well.

Serve with Broiled Butterfly Lamb.

RICE CARIBBEAN STYLE

preparation time: 5 minutes   *Silvia Kahn*
cooking time: 30 minutes   *Antipodes*
serves: 6

2 cups white Uncle Ben's rice
Salt and pepper
1 Tblsp. butter
1 Tblsp. olive oil
2 fresh garlic cloves

1 tsp. brown sugar
1 sliced white onion
Turmeric
Coriander
4 cups chicken stock (or cubes)

Saute the rice in oil and butter, stirring, 2 minutes. Add sugar, garlic and 4 cups of chicken stock and the rest of the ingredients. Slice onion and let it float on top for taste, then remove when done. Cook till boiling, then lower heat, cover and cook 20 minutes or till done. Press in a Pyrex dessert bowl and turn upside down on dinner plate like a mold form. Top with shallots and serve with hot chicken and pineapple.

COCONUT RICE

preparation time: 15 minutes   Nancy Thorne
cooking time: 20 minutes   Gullviva
serves: 6

1½ cups rice   2½ cups chicken stock
1 onion, chopped   ⅓ cup Coco Lopez (any cream of
½ cup shredded coconut   coconut)
1 Tblsp. curry powder   ½ tsp. salt
2 Tblsp. butter

Saute onion, coconut, rice, curry powder till coconut is lightly toasted.
Add stock, cream of coconut and salt and simmer till rice is cooked.

FRIED RICE

preparation time: 35 minutes   Barbara Haworth
cooking time: 10-15 minutes   Ann-Marie II
serves: 4

6 slices bacon   ½ cup water
½ cup scallions, chopped   4 eggs, slightly beaten
¼ cup soy sauce   2 cups cooked rice

Preheat wok or skillet and fry bacon. Drain and crumble. Stir fry onions
1-2 minutes in drippings. Add rice. Stir fry 7-8 minutes. Add soy sauce,
water and bacon. Reduce to low heat. Pour in eggs on top. Stir fry 3-4
minutes or till eggs are cooked.

RICE MALIA

preparation time: 5 minutes   Betsi Dwyer
cooking time: 20-25 minutes   Malia
serves: 6

2-3 cups cooked white rice (1 cup   ⅓ cup slivered almonds, toasted:
   uncooked)
3 heaped Tblsp. peach chutney

Add chutney and almonds to cooked rice. The key to perfect white rice:
simmer over the lowest possible heat for 15 minutes without removing
the pan lid (1 cup uncooked rice and 2 cups water).

LEMON RICE

preparation time: 15 minutes
cooking time: 15-20 minutes
serves: 6

Barbara Haworth
Ann-Marie II

⅔ cup Minute rice
⅔ cup water
¼ cup celery
2 Tblsp. onion, chopped
¼ cup mushrooms, chopped
2 Tblsp. margarine or oil

Dash thyme or blended herbs
¼ tsp. salt
Dash pepper
1 Tblsp. lemon juice
¼-½ tsp. grated lemon peel

Cook rice. Saute celery, onion and mushrooms in margarine. Mix with rice and add remaining ingredients. Heat on low heat till hot.

LEMON RICE FOR SEAFOOD

preparation time: 5 minutes
cooking time: 15 minutes
serves: 6

Shirley Fortune
Maranatha

1 large lemon
1½ cups rice

3 cups water
2 tsp. salt

Grate lemon rind or use zester and set aside. Squeeze lemon and combine three tablespoons of juice with the rice, water, and salt in your pan. Bring to a boil and simmer 15 minutes. *Serve rice garnished with lemon rind. For use with scampi, add rind to water for zippier flavor.*

WHITE RICE WITH ALMONDS

preparation time: 2 minutes
cooking time: 20 minutes
serves: 12

Paula Taylor
Viking Maiden

2 cups Uncle Ben's rice
5 cups water
2 Tblsp. butter

Salt
4½ oz. slivered almonds

Cook rice as per instructions on pkg. Add slivered almonds 5 min. before rice is cooked.

WEST INDIAN RICE

preparation time: 10 minutes   *Katie MacDonald*
cooking time: 30 minutes   *Oklahoma Crude II*
serves: 6

Rice and water (to make about 4   1 tsp. basil
   cups cooked)   2 Tblsp. soy sauce
1 Tblsp. curry   Dash Tabasco
4 cloves garlic, crushed   Pepper
1 onion, chopped   Salt
¼ cup butter   1 egg

Cook rice as you normally would, including all ingredients, except for
egg. Just before rice is cooked, push rice to sides of pan and scramble
egg in center of pan gradually mixing with rice. *Serve immediately.*

CALYPSO RICE

preparation time: 10 minutes   *Kandy Popkes*
cooking time: 20 minutes   *Wind Song Oregon*
serves: 6

2 Tblsp. butter or margarine   1 cup uncooked rice
1-2 tsp. curry powder   ⅓ cup raisins
1 Golden Delicious apple, cored,   2 cups apple juice
   coarsely chopped   1 tsp. salt
½ cup chopped onion   ½ cup dry roasted cashew nuts,
½ cup sliced celery   halved

Melt butter or margarine in large frying pan. Stir in curry powder; add
chopped apple and cook slowly 2 minutes, stirring constantly. Remove
apple and set aside. Add onions and celery to frying pan and cook 1
minute, stirring constantly. Add rice, raisins, apple juice and salt. Bring to
a boil. Reduce heat, cover tightly and simmer 15 minutes or until rice is
tender and liquid is absorbed. Gently stir in apple and nuts.

RICE AND CARROT PILAF

preparation time: *5 minutes*
cooking time: *15 minutes*
serves: *4*

Launa Cable
Antiquity

1 can (13¾ oz.) chicken broth
⅓ cup water
2 cups enriched rice, precooked

2 large carrots, grated (about 1 cup)

Combine chicken broth and water in medium-sized saucepan, bring to a boil. Stir in rice and carrots. Cover and remove from heat. Let stand 5 minutes. Fluff with fork.

MEATS

SUGGESTED MENUS

COLD MINTED PEA SOUP
VEAL WITH ARTICHOKES AND LEMON SAUCE
WHITE RICE WITH ALMONDS
HARICOTS VERTS PROVENCALE
FRENCH CHOCOLATE CAKE

PEACH SALAD
STEAK AU POIVRE
HASSELBACK POTATOES
BROCCOLI IN LEMON SAUCE
AMARETTO CREAM

CHILLED CUCUMBER SALAD
FRUIT STUFFED LEG OF LAMB
NUTTY WILD RICE
PETIT POIS PEAS AND SCALLIONS
NEW ZEALAND PAVLOVA

BEEF WELLINGTON

preparation time: 45 minutes
cooking time: 40 minutes plus 25
serves: 8

Terry Boudreau
High Barbaree

3½ lbs. fillet of beef
2 Tblsp. brandy
½ tsp. salt
½ tsp. pepper
½ tsp. Accent
6 slices bacon
½ lb. finely diced mushrooms

4 Tblsp. soft butter
½ cup liver paté
1 beaten egg
1 pressed garlic clove
3 Tblsp. chopped parsley
1 pastry recipe for 2 crust pie

In morning, preheat oven to 325°. Place beef in shallow roasting pan. Rub with brandy, salt and pepper, Accent, butter and garlic. Top with bacon. Roast 10 minutes per pound. Remove from oven to cold platter and cool immediately. Sauté mushrooms in butter. Add pàté and parsley. Roll pastry in oval big enough to wrap beef completely. Spread with mushroom mixture. Place fillet on pastry. Wrap pastry around it, trimming and tucking in edges. Brush with beaten egg. Make diagonal slits in pastry. Decorate with flowers cut from pastry. Bake in 425° oven for 25 minutes.

Serve Artichokes with Lemon Butter, Summer Squash Casserole, Golden Curried Potatoes and Orange Sauce Crepes.

FILLET OF BEEF L'ORLY

preparation time: 30 minutes
cooking time: 1 hour
serves: 8-10

Bobbie Gibson
Gibson Girl

8 lbs. fillet of beef
½ cup butter, softened
Salt and pepper, to taste
Garlic salt, to taste

2 Tblsp. Worcestershire sauce
½ cup bottled catsup
½ cup water

Bring fillet to room temperature, do not lard it (eye of round may be used but not as good). Rub with butter, seasonings and Worcestershire. Place in shallow roasting pan under broiler 15 minutes per side. Set aside until 1 hour before serving. (Go scrub your bottom (boat) or manicure your nails. Anything foolish). Roast at 375° for 30-45 minutes (30 min. for rare). 15 minutes before done, remove drippings from pan to saucepan. Add catsup and water, simmer 5 minutes. Pour mixture over beef and roast 10 more minutes. *Serve with sauce.*

BEEF WELLINGTON A LA LINDA

preparation time: 30-40 minutes
cooking time: 1¾ hours
serves: 8-10

Linda Richards
Tao Mu

5 lb. boneless top loin roast
1 (4 oz.) can deviled ham
2 Tblsp. fresh minced onion (dry
   onion may be substituted)
½ tsp. hot pepper sauce
½ cup sour cream
1 clove garlic
Salt and pepper, to taste

Rich Pastry:

3 cups flour
1 tsp. salt
½ cup margarine
Cold water to form a stiff but
   workable pastry.

Make pastry and chill while preparing the meat. Mix all ingredients (except meat) into a thick paste. Rub meat on all sides with split clove of garlic. Spread paste on 3 sides of meat. Roll pastry until fairly thin, but easy to handle, reserving ⅓-½ cup for decoration. Spread paste on center of pastry the width of the meat. Place meat in the center of the pastry and enclose — sealing edges neatly. Decorate with remaining dough. This is your chance to be artistic. Brush tops and sides with beaten egg and bake on rack in roasting pan in 350° oven. *Serve Broccoli with Cheese Sauce, a tossed green salad, and Cherry Tarts for dessert.*

BEEF A LA BAREFOOT

preparation time: 15 minutes
cooking time: 5 hours
serves: 6

Phoebe Cole
Barefoot

3 lbs. sirloin
1 can mushroom soup
1 soup can of red wine

1 pkg. Lipton Onion Soup Mix
Pearl onions
Fresh mushrooms

Cut sirloin into 1-inch cubes. Combine all ingredients and cook in slow oven (250°) for 5 hours or more. Stir occasionally. Flavor improves with time. *Serve with French bread, salad with Anchovy Dressing and followed by Bananas Barefoot.*

BEEF BURGUNDY

preparation time: 25 minutes
cooking time: 2 hours
serves: 6

Kandy Popkes
Wind Song Oregon

6 strips bacon, cut in squares
3 lbs. beef chuck, cut in cubes
1 lg. carrot, sliced
1 can beef broth, undiluted
4 Tblsp. flour
1 tsp. salt
¼ tsp. pepper

2 cloves garlic, mashed
½ tsp. thyme
1 bay leaf
2 cups burgundy wine
1 (6 oz.) can tomato paste
½ lb. mushrooms
1 medium onion

Saute bacon. Brown beef in drippings. Saute carrots and garlic, add broth. Blend flour and seasonings in wine. Combine meat, carrots and wine mixture. Add tomato paste, blend and simmer 2 hours or until tender. Add mushrooms and onion last 30 minutes.

SAVORY PEPPER STEAK

preparation time: 30 minutes
cooking time: 1½ hours
serves: 6

Erica Benjamin
Caribe Monique

1½ lbs. round steak, cut ½ inch
    thick
¼ cup all-purpose flour
½ tsp. salt
⅛ tsp. pepper
¼ cup cooking oil or shortening
1 (16 oz.) can tomatoes
1¾ cups water

½ cup chopped onion
1 small clove garlic, minced
1 Tblsp. beef-flavored gravy base
1½ tsp. Worcestershire sauce
2 large green peppers, cut in
    strips
Hot cooked rice

Cut steak in strips. Combine flour, salt and pepper; coat meat strips. In large skillet cook meat in hot oil till browned on all sides. Drain tomatoes, reserving liquid. Add tomato liquid, water, onion, garlic and gravy base to meat in skillet. Cover and simmer for about 1¼ hours, till meat is tender. Uncover, stir in Worcestershire. Add green pepper strips. Cover and simmer for 5 minutes. If necessary, thicken gravy with a mixture of 1 Tblsp. cornstarch blended with cold water. Add drained tomatoes; cook about 5 more minutes. *Serve over hot rice.*

BEEF & BROCCOLI IN OYSTER SAUCE

preparation time: 10 minutes
cooking time: 15 minutes
serves: 6

· Abbie T. Boody
Ductmate I

2 lbs. flank steak, sliced thinly
1 bunch broccoli flowerettes
2 oz. minced ginger

1 bottle oyster sauce
1 oz. oil

Stir-fry ginger and beef in hot oil. Add remaining ingredients and cook 10 minutes. *Serve over white rice.*

BEEF STIFADHO

preparation time: 45 minutes
cooking time: 2-3 hours
serves: 8

Jane Marriott
Jan Pamela II

3 Tblsp. olive oil
32 tiny white onions
3½ lbs. stewing beef
6 cloves garlic (you don't have to use this many — I tend to have a heavy hand with the garlic)
3 Tblsp. flour
1 small can tomato paste
1 large can Italian plum tomatoes, drained

2 cups red wine
½ cup strong beef stock
"Kitchen Bouquet", salt and pepper to taste
1 Tblsp. ground cinnamon
1 tsp. freshly-grated nutmeg
2 tsp. ground coriander
3 bay leaves
½ cup raisins

First, peel the onions by dropping them in boiling water for a few minutes. You will then find that the skins slip off more easily, and that the operation is not so much of a tear-jerker as it might otherwise be. Next, heat the oil in a large sauce pan, and brown the onions over fairly fierce heat, stirring them often. Cut the meat into chunks and shake around in a plastic bag with the flour. Add to the pan, together with the garlic, and brown the meat on all sides. Then simply add everything else, bring to the boil, put lid on and simmer until cooked (2-3 hours, depending on the quality of the meat). *Serve with small, plainly boiled potatoes, macaroni or rice.*

DEBBIE'S RIBS

preparation time: 10 minutes
cooking time: 1½ hours
serves: 6

Maureen Stone
Cantamar IV

1 lb. ribs per person
1 cup ketchup
½ cup liquid honey

¼ cup soy sauce
5 cloves garlic, crushed

Simmer ribs in large pan of water for 45 minutes. Drain. Combine all ingredients. Pour over ribs in baking pan and bake for 45 minutes at 400° turning occasionally.

DAD'S B-B-Q SPARE RIBS

preparation time: 1 hour
cooking time: 3-4 hours
serves: 12

Paula Taylor
Viking Maiden

6 cups sliced onions
6 cups ketchup
6 cups water
½ cup salt
⅔ cup Worcestershire sauce

3 cups vinegar
3 cups brown sugar
½ cup dry mustard
¼ cup paprika
12 lbs. spare ribs

Mix all ingredients, except ribs, together in a large bowl. Cut spare ribs and braise to seal juices. Place ribs in two large, deep disposable pans. Pour sauce over ribs and cook at 350° for 3-4 hours, the longer and slower these cook the better they are. If ribs seem to be cooking too fast, lower heat in oven. Ribs are done when pans are a mess and ribs on top are starting to look a little overcooked.

SIRLOIN STRIPS A LA BREZINSKI

preparation time: 10 minutes
cooking time: 5-10 minutes
serves: 6

Betsi Dwyer
Malia

6 (8 oz.) strip steaks
1 cup barbecue sauce

6 Tblsp. garlic salt
Salt and pepper, to taste

A charter guest who is a meat market owner suggested this as a quick and easy way to enhance barbecued meats. Tenderize steaks, then add salt, pepper and rub in 1 Tblsp. barbecue sauce into each steak. Grill, broil, barbecue to desired doneness. *Serve with Potatoes Au Gratin and Grated Zucchini, followed by Dump Cake. Fruit Salad with Lime Sauce as an appetizer.*

FOOL PROOF BARBECUED STEAK

preparation time: 5 minutes   *Jean Thayer*
cooking time: 40 minutes   *Paradise*
serves: 6

4½ lb. cross rib roast, cut in half,   8 Tblsp. butter
   each half approximately 2   2 cloves garlic, halved
   inches thick   1 tsp. Worcestershire
4 Tblsp. lemon pepper

Steak must be at room temperature and barbecue coals flaming hot. Rub
lemon pepper into both sides of steaks. Place steaks 3" above coals and
grill 6 minutes each side. Put grilled meat on a heat-proof platter and
place in a pre-heated 300° oven for 15 minutes. Remove from oven and
reduce temperature to 250°. Rub each side of steak with cut cloves of
garlic and 6 Tblsp. of butter. Put remaining 2 Tblsp. of butter on top of
steaks and sprinkle with Worcestershire. Return steak to oven for 10 minutes. Remove steaks from oven and let rest for 5 minutes before carving
into thin slices. Accumulated juices may be passed separately. The steak
will be uniformly rare. *Serve with Grilled Red Onions, Baked Potatoes,
salad and Mt. Gay Bananas.*

ROAST BEEF PROVENCALE

preparation time: 20 minutes   *Jennifer Royle*
cooking time: 1½-2¼ hours   *Tri-World*
serves: 8-10

5-6 lbs. boneless rib eye   *Brown gravy:*
2 cups fresh white breadcrumbs   2 Tblsp. meat drippings
½ cup chopped parsley   3 Tblsp. flour
2 cloves garlic, crushed   1 can (10½ oz.) beef broth,
2 tsp. salt   undiluted
½ tsp. pepper   ¼ tsp. salt
½ cup butter   Dash pepper
4 Tblsp. Dijon mustard

Preheat oven to 325°. Combine breadcrumbs, parsley, garlic, 2 tsp. salt
and ½ tsp. pepper. Spread mustard over meat and pat crumb mixture
firmly into mustard. Drizzle with butter. Roast until degree of cooking desired. Remove roast. Make brown gravy: Pour off drippings in roasting
pan. Return 2 Tblsp. drippings to pan. Stir in flour; brown over low heat.
Add water to beef broth to make 2 cups, slowly add to flour mixture.
Bring to boiling and stir until smooth and bubbly. *Serve with whipped potatoes, cauliflower and cherry tomatoes stuffed with horseradish. Appetizer: Parsnip, Potato and Leek Soup.*

STEAK AU POIVRE

preparation time: 10 minutes
cooking time: 5-10 minutes
serves: 4

Cheryl Anne Fowler
Vanity

4 sirloin steaks
¼ cup green peppercorns
2 Tblsp. butter

2 Tblsp. oil
¼ cup Cognac
1 cup heavy cream

Heat butter and oil in pan. Roll steak in peppercorns. Brown for 2 to 5 minutes on each side, according to taste. Drain and place steaks on heated platter. Discard any fat drippings and add Cognac to pan and deglaze remaining drippings. Heat pan and light Cognac. Pour cream into pan along with any juice from steaks and allow mixture to simmer and reduce. Top steaks with sauce and serve.

STEAKS WITH TARRAGON BUTTER

preparation time: 5 minutes
chilling time: 1 hour
cooking time: 5-10 minutes
serves: 6

Kathleen McNally
Orka

2 medium shallots
2 Tblsp. chopped fresh parsley
4 tsp. tarragon vinegar
½ tsp. crumbled dried tarragon

½ tsp. freshly ground pepper
½ cup (1 stick) butter, well chilled,
cut into small pieces
6 beef tenderloin steaks

Tarragon Butter:

Combine shallots, parsley, vinegar, tarragon and pepper in processor or blender and mince using several on/off turns. (I find mincing finely with a sharp French knife more than adequate.) Add butter and blend well. Transfer to wax paper and roll into cylinder. Refrigerate or freeze until firm.

Steaks:

Grill, broil, barbecue, or pan-fry steaks to desired doneness. Transfer to individual plates or large serving platter. Slice butter into 6 rounds. Set atop steaks. *Serve with Baked Potato Chips, Endive/Bacon and Pecan Salad. For dessert, Hot Brownie Souffle with Vanilla Ice Cream Sauce. Appetizer, Smoked Trout Mousse with Rye Heart Toasts.*

SURE FIRE RIB ROAST

preparation time: 10 minutes   *Roz Ferneding*
cooking time: 5-6 hours   *Whisker*
serves: 6-8

On Antares I always had a 20 lb. Rib Roast, on Whisker it's 10 lbs. No
matter what weight the rib roast, this is the only sure fire method. No
other type of roast will do.

Preheat oven to 350°.

Put completely thawed out roast in oven at 1 o'clock in the afternoon,
bake for exactly one hour. Turn off oven and DO NOT OPEN OVEN
DOOR ALL AFTERNOON. I put my foil wrapped potatoes in at the
same time. ¾ of an hour before serving, turn oven ON to 350° again.
After you take the roast out of the oven, let it sit for 10 minutes before
carving.

This method works for any size Rib Roast ONLY, and will come out medi-
um-rare.

GREEK MEAT BALLS

preparation time: 20 minutes
cooking time: 30 minutes   *Saracen*
serves: 6

1 garlic, minced	**¼ cup chopped pine nuts**
2 lbs. lean ground beef	**¼ cup chopped onion**
½ cup lightly packed fresh mint	**1½ tsp. salt**
½ cup fresh parsley	**Flour**

Combine ingredients except flour and form balls (12). Roll balls in flour.
Freeze. Grill over high heat until brown outside and rare inside. 30 minutes before serving, heat oven to 300°. Bake for 25 to 30 minutes until
done as desired.

VEAL MARSALA

preparation time: 30 minutes   *Ann Doubilet*
cooking time: 20 minutes   *Spice of Life*
serves: 8

8 veal scallops   **¼ cup olive oil**
Flour, seasoned   **½ cup Marsala wine**
¼-½ lb. butter   **¼ cup chopped parsley**

Pound the scallops between sheets of wax paper until very thin. Cut into
1½-inch strips and dredge lightly with flour on one side only. Heat ½
butter and ½ olive oil and add scallops, flour side down first. Do not
crowd. You will need to do several batches. Cook about 3 minutes on
each side until tender and done. Remove to platter and keep warm. Add
butter and olive oil to pan as needed. When all the veal is cooked, deglaze the pan with Marsala, add parsley and a little butter, if needed to
taste. This is a quick dish to prepare, especially if you pound the veal
ahead of time. *Serve with fettuccini noodles made in Italy, mixed with
fresh grated parmesan, crusty hot French bread, and a green salad.*

VEAL SCALLOPS MARSALA

preparation time: 15 minutes   *Chris Kling*
cooking time: 15 minutes   *Sunrise*
serves: 6

6 veal scallopine cutlets   **1 cup mushrooms, quartered**
⅓ cup flour   **2 lemons**
1 tsp. salt   **¼ cup Marsala wine**
¼ tsp. pepper   **Chopped parsley**
¼ lb. butter

Clean quarter mushrooms. Melt butter in a large heavy skillet. Saute
mushrooms in butter for 3 to 5 minutes. Remove with a slotted spoon.
Combine flour, salt and pepper and dredge veal cutlets in the mixture.
Lightly brown the cutlets in the skillet. Add wine and juice of one lemon.
Cover and simmer for about 5 minutes. Add mushrooms and simmer 1
minute longer. Garnish with lemon slices and chopped parsley. *Serve with
a tossed green salad, Tomatoes Florentine and Manicotti. Amaretto
Cheesecake for dessert.*

Pernilla Stahle on Adventure III, sautes sliced shallots with the mushrooms and sprinkles parmesan cheese on top of the veal.

ELEPHANT STEW

preparation time: 2 months　　　　　　　　*Pamela McMichael*
cooking time: 4 weeks　　　　　　　　　　　　*Pieces of Eight*
serves: 3800

1 large elephant　　　　　　　　**Salt and pepper**
Brown gravy　　　　　　　　　　**2 rabbits**

Cut the elephant into bite-sized pieces. (This should take about 2 months.) Add brown gravy to cover and salt and pepper. Cook over a kerosene fire about 4 weeks at 465°. *Makes about 3800 servings. If more guests are expected, add 2 rabbits. Do so only if necessary, however, as most people do not like HARE in their stew.*

VEAL PICATA

preparation time: 30 minutes　　　　　　　　*Candy Glover*
cooking time: 30 minutes　　　　　　　　　　　*Anemone*
serves: 4

2½ lbs. veal cutlets　　　　　　　**4-8 oz. unsalted butter**
½ cup seasoned breadcrumbs　　　**Capers**

Pound veal with meat mallet until veal is thin and well tenderized. Pieces should be about 3 x 8. Lightly bread veal. In large skillet melt 4 oz. of unsalted butter. On high heat cook 4 to 5 pieces at a time, turning only once, 3-4 minutes. Sprinkle with capers as you turn them. Place veal in top of double boiler to keep warm as you finish cooking the rest of the veal. Add additional butter as you need it. Garnish with lemon wheels.

Roz Ferneding on Whisker cooks the veal (uncrumbled) in butter and lemon juice. Removes the veal, adds lots more butter, a little more lemon juice and 1 cup of chicken stock. Brings it all to the boil, adding flour and stirring until thick. Serves with spaghetti lightly buttered.

VEAL DIJON

preparation time: 15-20 minutes   Betsi Dwyer
cooking time: 15-20 minutes   Malia
serves: 6

2 Tblsp. vegetable oil   ⅓ cup dry vermouth
4 Tblsp. sweet butter   ⅓ cup Dijon mustard
3 scallions, chopped   ½ cup heavy cream
9 veal cutlets, dredged in flour   Pimientos
Salt, to taste   Dill
Pepper, to taste

Melt the butter and oil in a large skillet. Add the scallions and cook over low heat for a few minutes. Raise the heat and add the veal, seasoned to taste. Cook for 2 minutes per side. Keep cooked veal warm. Add the vermouth to the skillet and bring to a boil. Cook until mixture is reduced to a few thick spoonfuls. Whisk in the mustard and cream and boil for 2 minutes. *Serve over a bed of hot noodles on a platter and surround with Tomatoes Provencale. Garnish platter with pimiento pieces and a few sprinkles of dill.*

VEAL IN CREAMY WINE SAUCE

preparation time: 20 minutes   Alison P. Briscoe
cooking time: 30 minutes   Ovation
serves: 10

10 veal cutlets   3-4 Tblsp. margarine or butter
1 onion, finely chopped   1 soup can white wine (medium
1 green pepper, finely chopped   dry)
2 cans cream of chicken soup   ½ tsp. thyme
2 small cartons whipping cream   Salt
¼ lb. chopped, fresh mushrooms   Freshly ground black pepper
Seasoned all-purpose flour   ¼ tsp. paprika

Dip veal in seasoned flour and saute in margarine or butter for 5 minutes. Remove veal, place in baking dish, saute onion, pepper and mushrooms lightly. Place onion, pepper, mushrooms in baking dish. Mix cream, soup and wine and all seasonings and cover ingredients in dish. Cook at 350° (preheated) for 30 minutes. Check seasonings after 20 minutes. *Serve with a mixture of brown and wild rice, tossed green salad, steamed fresh green beans.*

VEAL SENTINO

preparation time: 30 minutes   *Lori Moreau*
cooking time: 15 minutes   *Alberta Rose*
serves: 6

6 veal scallops, flattened   **4 oz. butter**
1 tall can asparagus, drained   **6 oz. Swiss cheese, grated**
½ lb. mushrooms, sliced   **Tarragon**
Juice of 1 lemon

Saute mushrooms in 2 oz. butter. Set aside. Saute veal very lightly on both sides. Remove to a baking dish, placing in a single layer. Sprinkle with tarragon. Arrange asparagus spears on top of veal (approximately 4 to each). Place mushrooms on top of asparagus. Cover with cheese. Melt remaining butter with lemon juice and pour over all. Bake, covered, at 375° for 15 minutes. *Serve with Fettuccine Gene, Italian Salad and Wine Cake.*

VEAL WITH ARTICHOKES AND LEMON SAUCE

preparation time: 20 minutes   *Jan Robinson*
cooking time: 30 minutes   *Vanity*
serves: 4

8 veal scallopine (¼ inch thick)   **14 oz. can artichoke bottoms**
Flour to coat   **½ cup dry white wine**
Pepper   **1¼ cups chicken bouillon or stock**
½ stick butter   **2 lemons**
¼ cup finely chopped shallots or   **½ cup heavy cream**
**   onion**

Heat butter in electric skillet or large frying pan. Coat veal in flour and pepper. Saute until golden brown, turning once. Add chopped shallots, drained artichoke bottoms and pour in the white wine. Bring the mixture to a simmer. Add enough stock to cover veal completely. Grate lemon rinds and set aside. Add squeezed lemon juice to sauce. Cover the skillet and cook over low heat for 20 minutes. Stir in the cream and simmer for 5 minutes uncovered. Garnish on top of veal with grated lemon rind. *Serve with buttered noodles and a green vegetable.*

VEAL WITH GINGER WINE SAUCE

preparation time: 15 minutes
cooking time: 15 minutes
serves: 4

Guy Howe
Goodbye Charlie

4-6 veal scallops
½ cup flour
pepper and salt
½ cup butter

1 Tblsp. olive oil
1 cup Stones Green Ginger Wine
¼ cup lemon juice
½ cup heavy cream

Dust veal with flour, pepper and salt. Heat butter and oil in skillet and saute veal until brown on both sides. Remove and arrange on warm serving dish and keep warm in oven. Add remaining flour to skillet and brown. Add ginger wine gradually. Bring to boil, reduce heat and simmer about 5 minutes. Stir in lemon juice and cream and reheat. Season to taste. Pour over veal. Garnish with parsley and lemon twists.

CREOLE ROAST LAMB

preparation time: 10 minutes
cooking time: approximately 2 hours
serves: 6-8

Margaret Benjamin
Illusion II

1 leg of lamb (5-6 lbs.)
1 Tblsp. wine vinegar
2 Tblsp. Worcestershire sauce
2 tsp. chopped thyme
½ bay leaf, crushed

Dash Tabasco
Salt and pepper
½ pt. boiling stock (chicken or beef)
2 grated onions
2 cloves garlic

Make small slits in the skin and insert slices of garlic from one garlic clove. Roast lamb in oven at 350°. Meanwhile, mix all other ingredients together and baste the meat with it occasionally. When the lamb is cooked, remove from roasting pan, pour in the rest of the sauce and heat. Carve lamb and pour sauce over. *Serve with roast potatoes, eggplant and tomato ragout and Lemon Crunch Pie. Appetizer: Melon Salad.*

FRUIT STUFFED LEG OF LAMB

preparation time: 40 minutes　　　　　　　　　　*Casey Wood*
cooking time: 2½ hours　　　　　　　　　　　　　*Megan Jaye*
serves: 6-8

11 oz. dried mixed fruit　　　　　　**½ tsp. nutmeg**
2 cups orange juice　　　　　　　　**4-5 lb. boneless leg of lamb**
1 cup white wine　　　　　　　　　**Salt**
2 cinnamon sticks　　　　　　　　　**Pepper**
½ tsp. ginger　　　　　　　　　　　**Orange juice**

Heat dried fruit, 2 cups orange juice, 1 cup white wine, cinnamon, ginger and nutmeg to boiling. Reduce. Simmer 20 minutes. Cool. Discard cinnamon sticks. Lay roast flat, sprinkle with salt and pepper. Spoon about ⅓ of the fruit on meat. Roll and tie at intervals. Roast in shallow pan uncovered at 325° about 2 hours. Spoon remaining fruit mixture around meat during last 30 minutes adding more orange juice for desired consistency. *Serve sauce over sliced meat.*

BROILED BUTTERFLIED LAMB

preparation time: 20 minutes　　　　　　　　*Ann Doubilet*
marinating time: 12-24 hours　　　　　　　　*Spice of Life*
cooking time: 30 minutes
serves: 6-8

6 lb. boneless leg of lamb

Marinade:

⅔ cup olive oil　　　　　　　　**1 tsp. oregano**
Juice of 2 lemons　　　　　　　**3 bay leaves**
Salt　　　　　　　　　　　　　**1 large onion, sliced very thin**
Pepper　　　　　　　　　　　　**2 cloves garlic, sliced thin**
2 Tblsp. chopped parsley

Mix all ingredients of marinade in large baking dish. Open lamb up so it is a flat though rugged piece. Place meat in marinade, turning to coat. Marinate 12-24 hours, turning meat every 2-3 hours. Heat grill. Place meat fat side towards heat with marinade still coating it. Sprinkle with salt. Cook 15 minutes and turn, pour some marinade over it. Cook 15 minutes and test thickest part which should be rare. The pieces of meat are lumpy so you will have sections of rare, medium, and well-done. Slice as you would a large steak.

Serve with Egg and Lemon Sauce, Rice and Mediterranean Vegetables.

ROAST LAMB WITH CURRANT JELLY

preparation time: 20 minutes   *Ann Doubilet*
cooking time: 2 hours   ·   *Spice of Life*
serves: 8

5 lb. boneless or bone in leg of   **2 Tblsp. flour**
    lamb   **2 Tblsp. red currant jelly**

Preheat oven to 350°. Pat dry leg of lamb. Place in baking pan. Pat flour
over lamb. Place in oven about 15 minutes until flour is almost absorbed.
Remove from oven and spread currant jelly over surface. Return to oven
and bake 20 minutes per pound. *Serve with peas, new potatoes, and
tossed salad. Rum and Grand Marnier Chocolate Mousse for dessert.*

CURRY LAMB WITH RICE

preparation time: 30 minutes   *Pamela McMichael*
cooking time: 20 minutes   *Pieces of Eight*
serves: 12

4 cups boiling water   **4 Tblsp. onion, chopped**
6¼ lbs. leg or shoulder of lamb   **1 cup celery, chopped**
8-10 Tblsp. vegetable oil   **½ cup parsley, snipped**
2 tsp. salt   **6-8 Tblsp. flour**
1 tsp. pepper   **½ cup raisins**
5 or more tsp. curry powder

Trim away gristle and fat from lamb. Cut into 1-inch cubes. Brown in oil.
Drain. Add water, seasonings, onion, celery, parsley, and raisins. Cover
and simmer about 20 minutes or until done. Stir often. Thicken stock
with flour blended with small amounts of water. Cook and stir several
minutes longer. *Serve over rice. Garnish with chutney, slivered almonds,
coconut, and crumbled bacon.*

SKEWERED LAMB AND VEGETABLES

preparation time: 45 minutes
cooking time: 20 minutes
serves: 4

Pamela McMichael
Pieces of Eight

½ cup vegetable oil
¼ cup fresh lime juice
1 tsp. ground cumin
½ tsp. ground coriander
½ tsp. salt
½ tsp. pepper
3 scallions, crushed

3 garlic cloves, crushed
2 medium zucchini
1½ lbs. lamb shoulder, cut into
    1½" cubes
1 pint cherry tomatoes
Lime slices (optional)

Prepare hot coals for grill. Whisk oil, lime juice, cumin, coriander, salt and pepper. Stir in scallions and garlic. Cut zucchini into 1½-inch slices. Add zucchini and lamb to marinade. Let stand covered at room temperature 30 minutes, tossing occasionally. When coals are ready, place lamb and zucchini alternately on skewers. Place cherry tomatoes alternately with lime slices on separate skewers and set aside. Grill lamb skewers 4" above coals and turning and brushing once with marinade until lamb cubes are dark brown on outside and medium rare on inside, 18 to 20 minutes. 5 minutes before lamb skewers are done, place tomato skewers on grill and brush with marinade about 5 minutes. *Serve immediately with Bulgar Pilaf and Fruit with Orange Caramel Syrup for dessert.*

LAMB CHOPS A LA TREKKER

preparation time: 10 minutes
cooking time: 30 minutes
serves: 4

D. J. Parker
Trekker

8 center cut lamb chops
2 Tblsp. oil
2 apples, unpeeled, cored, sliced
2 Tblsp. brown sugar
1 tsp. cinnamon

1 Tblsp. cornstarch
½ cup water
Garnish:
Endive
Radish roses

In skillet, brown chops in oil. Cover, cook until tender. Remove chops to heated plate. Gently saute apples. Remove excess oil from pan. Mix together brown sugar, cinnamon, cornstarch, and water. Pour over apples. Stirring until thin sauce. Add more water only if necessary for consistency. Return chops to pan and cover with sauce. Cover pan and remove from heat for five minutes. *Serve hot with apples over chops. Garnish with endive and radish roses, Parmesan Zucchini Flares and parsley potatoes. Appetizer: Blue Cheese Salad. Dessert: Sailor's Coffee.*

STUFFED PORK CHOPS

preparation time: 30 minutes   *Erica Benjamin*
cooking time: 1 hour   *Caribe Monique*
serves: 6

6 (1¼-inch) pork chops   Salt
1½ cup toasted bread cubes   2 Tblsp. melted butter
½ cup cheddar cheese, shredded   2 Tblsp. orange juice
½ cooking apple, diced   Dash cinnamon
2 Tblsp. raisins

Cut pocket in pork chops and salt inside of pocket. Mix all ingredients except butter and juice. Melt butter and mix with juice and pour over mixture and mix well. Stuff inside pocket and bake in oven at 350°.

PORK ROAST WITH RUM SAUCE

preparation time: 15 minutes   *Jennifer Royle*
cooking time: 2½ hours   *Tri-World*
serves: 8

5-6 lbs. pork loin   1 medium bay leaf, crushed
2 cups chicken stock   1 tsp. salt
1 cup light brown sugar   ¼ tsp. pepper
2 Tblsp. dark rum   ¼ cup light rum
2 tsp. minced garlic   4 tsp. cornstarch blended with 2
2 tsp. ground ginger   Tblsp. cold water
½ tsp. ground cloves   3 Tblsp. lime juice

Score roast 1" apart fat side up. Roast at 350° for two (2) hours. Remove meat and skim fat from pan juices. Pour stock into juices and return meat to pan. Mash brown sugar, 2 Tblsp. rum, garlic, ginger, cloves, bay leaf, salt and pepper to smooth paste and spread over meat. Roast for additional 30 minutes. Remove meat. Warm ¼ cup light rum and ignite. Add lime juice. Thicken pan juices with cornstarch mixture. Remove from heat and add rum and lime mixture.

Serve with Chilled Zucchini Soup, fresh baked yams, green beans almondine and applesauce.

JEWELLED CROWN ROAST OF PORK WITH CHAMPAGNE DRESSING

preparation time: 10 minutes
cooking time: 30 minutes per lb.
serves: 8

Shirley Fortune
Maranatha

7 lb. crown roast of pork
3 Tblsp. lime juice
1 tsp. ground ginger
1 Tblsp. flour
1 tsp. celery salt
½ tsp. ground pepper

Parsley (garnish)
Green grapes in clusters
Small red crab apples
Kumquats, drained
Stuffing (recipe follows)

Pour lime juice over roast. Mix together ginger, flour, salt and pepper and sprinkle over meat. Roast at 350° watching carefully not to overcook. After one hour drain fat and fill with dressing. Return to oven and cover lightly with foil until done. *To serve: place pork on platter and surround with a wreath of parsley sprigs. Place bunches of green grapes on top of parsley. Small red crab apples are a pretty contrast between clusters. "Spear" kumquats on toothpicks on tops of ribs like jewels atop a crown. If kumquats are unavailable, substitute paper frills made from typing paper.*

Champagne Dressing:

preparation time: 30 minutes

1 cup chopped onion
½ cup crushed pineapple, drained
4 strips bacon, fried and crumbled
½ cup ripe olives, chopped
2 Tblsp. parsley, chopped
1 cup celery, chopped, sauteed 5
     minutes in butter

2 cups wild rice cooked in
     champagne or dry white wine,
     not water
1 tsp. marjoram
½ tsp. thyme
2 eggs, beaten
1 cup raw mushrooms, chopped

Combine all ingredients and stuff center of pork roast. Prepare after roast has been started and add after first hour of roasting.

B.B.Q. PORK CHOPS OR CHICKEN EASY

preparation time: 15 minutes   Roz Ferneding
cooking time: 1½ hours   Whisker
serves: 6

6 pork chops or chicken breasts   1 onion finely chopped
1½ cups pineapple juice   1 cup chili sauce
¼ cup brown sugar

Combine pineapple juice, brown sugar, onion and chili sauce. Pour over
chops or chicken. Bake covered 1 hour, uncovered ½ hour. If sauce is
too runny, add a little flour to thicken. Very easy.

PORK CHOPS WITH CRANBERRY-WALNUT STUFFING

preparation time: 40 minutes   Alberta Rose
cooking time: 1 hour
serves: 6

6 (1½-inch) pork chops   ⅜ grated orange rind
1½ cups herb stuffing mix   1 (16 oz.) can mushroom gravy
½ cup boiling water   ¾ cup orange juice
6 Tblsp. butter   ¼ tsp. cloves
¾ cup whole cranberry sauce   Cinnamon
⅜ cup walnuts

Combine stuffing mix, water and ½ the butter. Add the cranberry sauce,
walnuts, and orange rind. Mix gently. Make a slit in each chop and fill the
pocket with stuffing. Secure with picks. Brown on both sides in remaining
butter. Season with salt and pepper. Mix the gravy, orange juice and
cloves (thicken if desired). Chops may be simmered in a covered skillet or
baked, covered, at 325°. Either way, cooking time is 60 minutes. Before
simmering, pour gravy over chops and baste occasionally during cooking.
Halfway through, sprinkle with cinnamon, if desired. *Remove to serving
platter and garnish with orange twists and parsley, buttered baby carrots
and Sour Cream Raisin Pie. Easy Pumpkin Soup as the appetizer.*

PORK CHOPS FLAMED WITH BRANDY

preparation time: 15 minutes
cooking time: 45 minutes
serves: 8

Ann Doubilet
Spice of Life

8 or 16 pork chops
4 Tblsp. butter
2 shallots, minced

½ cup brandy
4 Tlbs. Dijon mustard
4 oz. sour cream

Melt butter in pan. Brown pork chops on both sides. Add shallots and brown. Remove from heat and add brandy. Swirl in pan and ignite. When flame dies, return to heat and add Dijon mustard. Place sour cream in cup and add some hot juice. Return mixture to pan, cover, cook about 15 more minutes until done. *Serve with broad egg noodles.*

SWEET AND SOUR PORK

preparation time: 35 minutes
cooking time: 15 minutes
serves: 6

Casey Wood
Megan Jaye

1-2 pork tenderloins
Soy sauce
Flour
16 oz. can or fresh pineapple
1 can sliced water chestnuts
1 can bamboo shoots
2 pkgs. Japanese vegetables
6 sliced sweet pickles

Salt
Pepper
Ginger
½ cup vinegar, tarragon, oil
Broccoli (optional)
Cauliflower (optional)
Mushrooms (optional)

Slice pork to ⅛ to ¼ inch. Dredge in soy sauce, then flour, brown in hot oil, add water chestnuts and bamboo shoots. Brown, reserving liquid. Add pineapple, reserve liquid. Add Japanese vegetables, pickles, all reserved liquids, and vinegar. Season to taste. Let simmer covered, 10-15 minutes. It is not necessary to add cornstarch to thicken the sauce; the flour will create it's own consistency. *Serve this one either over or next to fluffy white rice. Pass a bowl of Chinese noodles and some duck sauce, a tarragon marinated mushroom salad and eggrolls.*

SWEET AND SPICY PORK CHOPS

preparation time: 25 minutes
cooking time: 45 minutes
serves: 4

Liz Thomas
Raby Vaucluse

4 pork chops, ¾-inch thick
½ cup apricot/pineapple
    preserves
¼ cup soy sauce

¼ cup white wine
2 oz. oil and margarine mixed

Brown chops in hot fat. Combine other ingredients and pour over chops in shallow baking dish. Cover and bake at 350° for 45 minutes or until tender.

PORK TENDERLOIN WITH MUSTARD SAUCE

preparation time: 15 minutes
marinating time: 3 hours
cooking time: 1 hour
serves: 6

Jean Thayer
Paradise

2 Tblsp. brown sugar
¼ cup soy sauce

¼ cup rum — Mt. Gay
3 lbs. pork tenderloin

Mix ingredients and marinate pork several hours, turning occasionally. Preheat oven to 325°. Remove from marinade and bake 45 minutes to 1 hour, basting with marinade. *Carve into thin slices and serve with sauce.*

Mustard Sauce:

⅓ cup sour cream
⅓ cup mayonnaise

1½ tsp. dry mustard
2-3 green onions, chopped

Mix until smooth. *Serve the pork tenderloin with the mustard sauce and any remaining marinade, in separate bowls.*

LOUISIANA STYLE PORK CHOPS

preparation time: 30 minutes   Gay Thompson
cooking time: 45-60 minutes   *Satori*
serves: 8

11 pork chops (cut thick)	1 red onion, peeled and sliced thin
1½ cup uncooked rice	2 green peppers, sliced in rings
1 can consommé or beef broth	2 tomatoes, sliced
1 cup dry white wine	1 small jar capers

Brown pork chops on each side. Place in deep baking dish. Fill open
spaces between pork chops with rice. Add remaining pork chops and
once again, fill open spaces with rice. Separate onion rings and place side
by side on top of chops. Top this with green pepper rings and tomato
slices. Combine pan drippings, wine, consommé and enough water to
equal 3 cups liquid. Pour over rice. Sprinkle capers on top. Cover tightly
and bake at 350° for 45-60 minutes. This dish can also be prepared on
stove top, simmering 45-60 minutes or until rice is cooked.

*Serve with broccoli and almonds, Peach Salad and English Pounds. This
meal can be cut down to fit any group — ¼ cup dry rice and ½ cup liquid per person.*

HAM WITH COCONUT/RUM SAUCE

preparation time: 15 minutes   Ann Glenn
cooking time: 5 minutes   *Encore*
serves: 8

Sliced ham, Hormel Cure 81
Fruit for garnish

Coconut/Rum Sauce:

¼ cup cream of coconut (canned)	¼ cup mango chutney (can
¼ cup dark rum	substitute cran-orange relish
	or apricot jam)

Arrange overlapping slices of ham in 9 x 13 glass dish. Garnish with fruit
such as pineapple, dried apricot or prunes. Prepare Coconut/Rum sauce
by combining cream of coconut, rum and chutney in small skillet and stirring while heating. Use a portion of sauce to baste ham. Reserve rest to
serve warm in a small pitcher with a ladle. Micro the ham about 10 minutes, or until it is heated thoroughly.

*Serve with Corn Pudding and Spinach/Artichoke. Followed by Ruby Ba-
nanas, or Carrot Cake with Rum Cream.*

SIR FRANCIS DRAKE HAM BALLS

preparation time: 30 minutes
cooking time: 2 hours
serves: 6

Fran Bryson
Amoeba

1 lb. minced ham
1 lb. minced pork
2 cups breadcrumbs
1 cup milk (fresh or condensed)
2 eggs
1 tsp. salt (if needed, depends on
    ham)

Sauce:

1½ cup brown sugar
¾ cup vinegar
¾ cup water
1 Tblsp. dried mustard

Prepare sauce first. Simmer all sauce ingredients over a low heat until sugar is dissolved. Set aside until ham balls are prepared.

Combine ham, pork, breadcrumbs, milk, eggs, and salt. Form into balls and place in a flat dish and cover with the prepared sauce. Bake at 325° for 2 hours. Turn every 15 minutes for even browning. Serve with the sauce in a fondue dish. Freezes well.

POULTRY

SUGGESTED MENUS

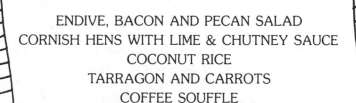

ENDIVE, BACON AND PECAN SALAD
CORNISH HENS WITH LIME & CHUTNEY SAUCE
COCONUT RICE
TARRAGON AND CARROTS
COFFEE SOUFFLE

CREAM OF PEANUT SOUP
EXOTIC GINGER CHICKEN
RICE CARIBBEAN STYLE
GREEN BEANS IN SOY BUTTER SAUCE
LEMON CRUNCH PIE

DIVINE APRICOTS
AWESOME CHICKEN
MANICOTTI
ZUCCHINI PESTO
COLD LOVE

AUTHENTIC PEKING DUCK

preparation time: 30 minutes
cooking time: 3 hours
serves: 4-6

Joel H. Jacobs
Bagheera

¼ cup honey
1½ dozen scallions
2 cloves garlic
1 cup plum sauce
2 - ¼" slices fresh ginger

1 prepackaged flour tortillas
2 Tblsp. salt
1 Tblsp. pepper
1 (5 lb.) Long Island Duck

Fill large size pan with water. Bring to a boil, add ¼ cup honey, two sticks sliced fresh ginger, ½ cup chopped-up scallions, one clove chopped garlic. Place cleaned duck into water and rotate 5 minutes until basted. Hang duck up to drain, preferably outside in the breeze. Use pan to catch drippings. When duck is dry rub outside with sliced garlic and season inside with salt and pepper, place in 350° oven. Cook for 20 to 30 minutes per pound or until skin is golden brown. Bone and cut duck into ½" by 2½" slices. Heat tortillas in 350° oven for 5 to 10 minutes. Remove and smooth on layer of plum sauce. Add duck slices and nicely trimmed scallions. Roll pancakes up and serve with Wan Fu wine. Use remaining plum sauce as dip. *Serve with steamed white rice, stir fried pea pods mixed with sliced water chestnuts fried in peanut oil and finely minced garlic, followed by Vanilla ice cream with black Bing cherries.*

CHERRY ALMOND DUCK

preparation time: 15 minutes
cooking time: 2 hours + 20 minutes
serves: 6-8

Marty Peet
Grumpy III

3 ducklings
16 oz. can Bing Cherries
½ cup slivered almonds

1 tsp. cinnamon
2 tsp. cornstarch, blended with a little cherry juice
salt and pepper

Preheat oven to 350°. Cut ducklings in half lengthwise, pierce skin, season with salt and pepper. Bake on a rack in a roasting pan 2 hours. Remove meat in chunks and pieces from carcass and place in baking dish. Save carcass for soup. Heat cherries, almonds, cinnamon and cornstarch mixture until slightly thickened. Pour mixture over duck pieces and heat in 350° oven for 15-20 minutes. *Excellent over wild rice.*

DUCK AU VIN L'ORANGE

preparation time: 30 minutes   Nancy Thorne
cooking time: 1¼ hours   Gullviva
serves: 6

2 ducklings
1 bottle red wine (3 cups)
10 black peppercorns
3 bay leaves
1 lemon, sliced
1 orange peel, julienne and
   blanched

1 small can orange juice
   concentrate
2 Tblsp. soy sauce
½ cup Grand Mariner
1 tsp. corn starch
Chopped parsley

Combine wine (preferably Pinot Noit, burgundy, etc.) peppercorns, bay leaves, sliced lemon and heat. Disjoint duck and place pieces over wine mixture and steam 30 minutes or until duck loses its pinkness. Probably will take two batches in steamer. (Vegetable steamer works well.) Remove duck and place in roasting pan. Cool wine mixture. Combine orange peel, orange concentrate, soy sauce, Grand Marnier and corn starch. Heat until thickened. Degrease wine mixture and add 1 cup to orange sauce. Bake duck 30 minutes at 350°. Baste often with orange sauce. Garnish with lots of fresh parsley. *Serve with Asparagus and Palms in Raspberry Vinaigrette, Nutty Wild Rice, Sauteed Red Onions and Carrots, and Eggnog Cheesecake.*

CREAMY CHICKEN ENCHILADAS

preparation time: 35 minutes
cooking time: 20 minutes   Saracen
serves: 2

2 Tblsp. butter
1 onion, thinly sliced
¾ cup shredded cooked chicken
2 Tblsp. diced green chilies
3 oz. cream cheese, diced chopped
   green onion

Sliced black olives & lime wedges
Salt
Oil
4 fl. tortillas
⅓ cup whipping cream
1 cup grated Monterey Jack

Grease pyrex dish. Melt butter in skillet & Saute onion. Remove from heat & add chicken, chilies, & cream cheese & salt. Mix. Lightly fry tortillas. Roll up enchiladas. Moisten top with cream and sprinkle with cheese. Bake at 375° uncovered about 20 min. *Garnish with green onion and olives. Squeeze lime juice over just before serving.*

GAME HENS WITH WILD RICE STUFFING

preparation time: 35 minutes
cooking time: 1 hour 15 minutes
serves 6

Chris Kling
Sunrise

6 Cornish Game Hens
1 (6 oz.) box Uncle Ben's Long
   Grain & Wild Rice
⅓ stick butter
2 Tblsp. minced onion
1 (2 ¾ oz.) can pate de foie with
   truffles

¼ cup cognac
2 Tblsp. minced parsley
1 egg yolk
Salt
Pepper
Kitchen bouquet

Preheat oven to 400° after preparing rice according to instructions on box using only 2 cups water. In a large skillet, saute onions in butter until transparent. Add cooked rice and the paté and stire until paté is dissolved. Add cognac and stir. Remove from heat and add parsley, egg yolk, salt and papper. Stuff each hen and rub outside of hens with kitchen bouquet. Roast for one hour and 15 minutes basting occasionally. Check for doneness on legs.
Serve with Asparagus and Nutmeg Butter, followed by Rum Chocolate Mousse. As an appetizer, French Onion Soup.

CHICKEN PERI PERI

preparation time: 30 minutes
cooking time: 1 hour
serves: 6

Shari Stump
Eyola

3 cornish hens
4 cloves garlic, pressed
1 onion, chopped
1 tsp. poultry seasoning

¼ lb. butter
1 tsp. Tabasco
Salt, to taste

Wash hens and pat dry with paper towel. Cut hens lengthwise on the underside, but not in two pieces, so the hens lay flat in "one piece" on its back. Press 1 clove of garlic and rub into hen. Use one clove per hen. Lightly salt and place in oven dish. Melt butter, add remaining garlic, onion, spices and Tabasco. Saute until onions are transparent. Pour sauce over hens and place in oven for 30 minutes at 350°. Turn over and cook for remaining 30 minutes, basting frequently. Serve over rice. *When hens are in the oven, be sure to give the cook a well deserved "Eyola's Cure All Bloody Mary."*

CORNISH HENS WITH OLIVES

preparation time: 20 minutes
cooking time: 1¼ hours
serves: 4

Jan Robinson
Vanity

2 medium onions
4-5 carrots
2 green peppers
3 stalks celery
6 Tblsp. butter
2 Tblsp. olive oil
4 Cornish Hens

Salt and pepper, to taste
1 cup chicken stock
½ cup dry white wine
1 (5 oz.) jar green olives
2 Tblsp. butter
4 Tblsp. flour

Peel, chop and saute onion, carrots, peppers and celery in the butter and olive oil. Preheat oven to 350°. Arrange sauteed vegetables on bottom of baking dish. Place hens on top of vegetables and pour chicken stock and wine over all. Roast uncovered for 1 hour, or until hens are cooked. Baste about every 20 minutes with juice from pan. When cooked remove to a warm platter. Place baking dish on top of stove to make the sauce with the vegetables. Add small amounts of beurre manie (2 Tblsp. butter and 4 Tblsp. flour kneaded together) cook and stir until slightly thickened. Add olives and heat through. Place hens on individual serving plates. Spoon vegetable sauce over hens. *Serve with a good stuffing. Garnish with a escarole or romaine lettuce leaf and a small bunch of purple grapes.*

HENS WITH LIME & CHUTNEY SAUCE

preparation time: 15 minutes
cooking time: 1 hour
serves: 6

Nancy Thorne
Gullviva

6 Cornish Hens
¼ lb. butter
2 Tblsp. curry powder
1 (12 oz.) jar Mango Chutney

Juice of 1 lime
1 medium papaya
Fresh parsley

Wash and dry hens. Melt butter and add curry. Paint birds with mixture and bake for 30 minutes at 350°. Mix chutney and lime juice. Paint birds with this mixture and bake 30 minutes more. Slice papaya and saute in pan juices. Serve, garnishing Cornish hens with papaya and parsley. *Serve with Fried Cheese, Coconut Rice, Broccoli with Almond Butter and Walnut Tart.*

BIRDS OF PARADISE

preparation time: 25 minutes
cooking time: 60 minutes
serves 6

Terry Boudreau
High Barbaree

6 Cornish Game Hens
Butter
Medium onion, chopped
1 cup chicken broth
1 cup white wine
1 Tblsp. parsley, chopped

3 minced garlic cloves
½ tsp. dried basil
1 bay leaf
½ tsp. thyme
Salt
Pepper

Season trussed birds with salt and pepper and brown on all sides in butter. Transfer to roasting pan. Saute onion for 5 minutes in pan where birds were browned. Add chicken broth, white wine, bay leaf, parsley, garlic, thyme, and basil. Bring to simmer and pour over roasting pan, cover and cook in a pre-heated oven (350°) for approximately 50 minutes. Remove cover and cook 10 minutes longer. *Serve on a bed of wild rice with garni of grilled tomatoes and Creamed Onions. Avocado and Grapefruit Salad and Strawberry-Orange Compote.*

MARINATED CORNISH HENS

preparation time: 20 minutes
marinating time: overnight
cooking time: 1 hour
serves: 6-8

Jan Robinson
Vanity

MARINADE:
4 cloves garlic, minced
1-⅓ cups lemon juice
1-⅓ cups dry wine vermouth
2 Tblsp. rosemary

1 tsp. salt, optional
2 tsp. black pepper
½ cup worcestershire sauce
6 Cornish Hens
1 cup dijon mustard

Combine ingredients for marinade and mix well. Wash hens and place in marinade for 24 hours in the refrigerator, turning 3 or 4 times throughout the marinating. Bring hens to room temperature. Brush hens inside and out with Dijon mustard. Place hens, breast up on a rack in the oven preheated to 450°. Reduce heat to 350° and bake for an hour, or until done. Baste hens several times with marinade while baking. *Serve hot, room temperature or chilled. They make a great part of a picnic lunch.*

LEMON ROSEMARY CORNISH HENS

preparation time: 15 minutes   *Launa Cable*
cooking time: 1 hour   *Antiquity*
serves 4

4 Cornish game hens   **2 Tblsp. lemon juice**
½ cup butter, melted   **½ tsp. Rosemary leaf, crumbled**

Rinse and dry hens. Combine melted butter, lemon juice and rosemary. Brush hens inside and out with mixture. Roast in oven for 50 minutes. Baste with remaining butter mixture and roast 10 minutes more or until juices run clear when thigh is pierced with fork.

DEVILED DRUMSTICKS

preparation time: 20 minutes   *Jean Thayer*
Marinating time: 1 hour   *Paradise*
cooking time: 35 minutes
serves: 6-8

4 lb. drumsticks, skinned   **2 tsp. thyme, dried**
1 cup sour cream   **2 tsp. dry mustard**
3 Tblsp. lemon juice   **1½ tsp. salt**
1 garlic clove, minced   **1 tsp. paprika**
Tabasco, salt, pepper, to taste   **¼ tsp. cayenne**
2 cups ground saltines   **¼ lb. butter, melted**

In large bowl combine sour cream, lemon juice, garlic, Tabasco, salt and pepper. Skin drumsticks and add to sour cream mix, tossing them to coat. Marinate for 1 hour at room temperature or chill overnight. In a shallow bowl, combine saltines, thyme, mustard, salt, paprika and cayenne. Roll drumsticks in crumb mix, shake of excess and arrange 1-inch apart on buttered baking sheet. Melt ¼ lb. butter and drizzle over drumsticks lightly. Bake at 375° for 35 minutes or till golden and tender. *Serve in a napkin-lined basket. Drumsticks may be hot or at room temperature. Excellent with Salad Provencal.*

STUFFING SUPREME

preparation time: 30 minutes
cooking time: according to weight of fowl
serves: 6

Carol Muller
"S'OUBLIER"

Herbal stuffing (Pepperidge Farm)
¼ lb. butter
1 cup water
1 cup celery
1 cup onion

1 golden or red apple
1 can pitted black olives
1 can artichoke hearts
½ cup green pepper

Melt butter in boiling water. Set aside. Into large bowl put 1 package herbal stuffing (7 pounds of poultry — 1 hen or 6 cornish hens). Add celery, onion, apple, pepper, all finely diced. Drain olives and artichoke hearts and add to mixture. Mix thoroughly with hands. Pour butter and water mix over all; mix thoroughly with hands. Stuff body and breast cavities of carefully cleaned fowl. Put in oven at once and bake at 350°.

MOM'S APPLE-SAUSAGE STUFFING
(for turkey 8-9 lbs.)

preparation time: 30 minutes
cooking time: 15 minutes
serves: 6-8

Cheryl Fleming
Silver Queen of Aspen

¾ lb ground sausage
2-3 medium size apples, cored and sliced into small cubes (skin on)
1 large onion, chopped fine

2 stalks celery, chopped fine
1 clove garlic, chopped fine
⅔ to ¾ bag Pepperidge Farm Seasoned Breadcrumbs
¼ lb. melted butter

Saute sausage for 10 minutes, discard grease, add apples, onion, celery and garlic. Saute until cooked. Combine all ingredients and let sit until ready to stuff bird. Cook bird as directed. Enjoy!

AWESOME CHICKEN

preparation time: 20 minutes
cooking time: 45 minutes
serves: 6

Betsi Dwyer
Malia

4 boneless chicken breasts, skinned and halved
16 oz. sour cream

2 cups seasoned bread crumbs
Butter

Preheat oven to 350°. Dredge chicken breasts in sour cream, then roll in bread crumbs. Place in baking dish, topping each breast with a pat of butter. Bake for 45 minutes or until tender. Dijon mustard can be substituted for the sour cream. *Serve with Hearts of Palm Tomato Salad w/ Eric's Dressing, Rice Malia, Tarragon and Carrots followed by Cay Lime Pie.*

NORTH CAROLINA CHICKEN

preparation time: 5 minutes
marinating time: 3-4 hours
cooking time: 1 hour
serves: 6

Jennifer Royle
Tri-World

5 lbs. chicken pieces
2 envelopes Italian Salad Dressing mix

½ cup lime juice
1 tsp. salt
8 oz. butter or margarine

Melt butter in saucepan. Stir in salad dressing mix, lime juice and salt. Marinate chicken 3-4 hours or overnight. Bake at 350° for 1 hour or until done. Or cook on outdoor grill for 1 hour turning and basting every 10-15 minutes.

CRISPY CHICKEN

preparation time: 15 minutes
cooking time: 1 hour
serves: 4

Roz Ferneding
Whisker

1 3 lb. chicken, cut in quarters
¼ cup evaporated milk
2 cups corn flake crumbs

Salt
Lemon Pepper

Dip chicken pieces in evaporated milk. Cover with corn flake crumbs. Place in baking pan. Sprinkle with salt and Lemon Pepper. Bake 1 hour at 350°.

LAZY MAN'S FRIED CHICKEN

preparation time: 15 minutes
cooking time: 1 hour
serves: 4

Launa Cable
Antiquity

2-3 lbs. frying chicken, cut up
¼ cup shortening
¼ cup butter
½ cup flour

1 tsp. salt
1 tsp. paprika
¼ tsp. pepper

Wash and dry chicken. In oven melt shortening and butter in 13x9x2 pan. Mix flour, salt, paprika and pepper. Coat chicken thoroughly with flour mixture. Place chicken, skin side down, in melted shortening. Bake uncovered for 30 minutes. Turn chicken and cook 30 minutes more.

FLYING JAKOB

preparation time: 20 minutes
cooking time: 20 minutes
serves: 6

Pernilla Stahle
Adventure III

4 grilled chickens
4 bananas
2 cups heavy cream
1 cup chile sauce

2 pkgs. bacon
1 cup peanuts
Italian salad seasoning
Salt and pepper

Clean chicken from bone. Put pieces in oven dish. Slice bananas lengthwise, put on top of chicken. Sprinkle with seasoning, salt and pepper. Whip cream and mix with chile sauce and pour it over chicken. Bake at 425° for 20 minutes. Garnish with crisp fried bacon bits and crushed peanuts. *Serve with Carrot Soup, Avocado/Corn salad, rice pilaf and Strawberry Fantasy.*

CHICKEN ALA ERICA

preparation time: 10 minutes
cooking time: 1 hour
serves: 4

Erica Benjamin
Caribe Monique

1 cut-up chicken
1 can cream of chicken soup

1 can cream of mushroom soup
1 can of heart of artichoke

Put chicken pieces in casserole dish. Cover with soups and artichokes and bake in a 325° oven uncovered.

Serve with mixed green salad and rice.

CHICKEN AND EGGPLANT

preparation time: 35-40 minutes   Barbara Haworth
cooking time: 25 minutes   Ann-Marie II
serves: 2-4

¾ lb. eggplant, julienne
2 Tblsp. lemon juice, plus water
   for boiling
1 cup chicken, julienne
2 Tblsp. soy sauce
1 tsp. brandy

2 Tblsp. cornflour
2 small green peppers, julienne
2 tsp. fresh ginger, grated
1 clove crushed garlic
½ cup chicken stock
¼-½ cup oil

Soak eggplant strips in boiling lemon water for 5 minutes. Drain. Set aside. Blend soy sauce, cornflour, ginger, and brandy. Coat chicken with mixture. Set aside. In wok or large skillet heat oil and garlic. Add pepper strips and fry 2 minutes. Set aside. Saute eggplant till tender, set aside. Saute chicken till tender. Combine all ingredients, cover and heat for 1 minute. *Serve with rice.*

AVOCADO & CHICKEN OR SHRIMP

preparation time: 45 minutes   Kandy Popkes
cooking time: 15 minutes   Wind Song Oregon
serves: 6

¼ cup butter (½ stick)
½ cup chopped apple
¼ cup chopped onion
1 clove garlic, crushed
1 Tblsp. curry powder
¼ cup flour
1 cup light cream
1 cup chicken bouillon

1 teaspoon salt
⅛ tsp. pepper
2 cups cooked chicken cut up or
   1½ lbs. shrimp, cooked,
   shelled & cleaned
3 avocados, halved & peeled
3 to 4 cups cooked rice

In saucepan: saute apple, onion, garlic and curry powder in the butter until onion is tender. Stir in flour, then gradually add cream and bouillon. Cook and stir until sauce boils and continue to boil for 1 minute. Add salt and pepper and chicken, or shrimp. Cook over low heat for 10 minutes. Arrange avocado halves on rice in heat proof serving dish. Heat in 350° oven for 5 minutes. Spoon curried chicken over avocado halves. *Serve with Indian or Euphrates bread.*

FRAN'S CHICKEN

preparation time: 10 minutes
cooking time: 40 minutes
serves: 4-6

Frances Bryson
Amoeba

5 lb chicken, cut-up
1 cup onions, chopped
5 cloves garlic, crushed
1 inch fresh ginger, crushed
4 cloves
4 Tblsp. Indian curry powder

2 sticks cinnamon
1 (16 oz.) can tomatoes
¼ pt. coconut cream
Salt
1 tsp. chili powder

Saute onions, garlic and ginger, add curry, chili powder and some juice from the tomatoes. Add cloves, cinnamon and chicken. Cook covered until chicken is tender, approximately 40 minutes. Stir in coconut cream just before serving.

CHICKEN & ARTICHOKE AU GRATIN

preparation time: 30 minutes
cooking time: 40 minutes
serves: 8

Kathleen McNally
Orka

8 boneless chicken breasts
Parsley
Salt
Pepper
Dash of garlic powder
3 cans artichoke hearts
1 cup mayonnaise

2 cans cream of asparagus soup
½ cup dry white wine (optional)
1 Tblsp. fresh lemon juice
1 cup shredded sharp cheddar
   cheese
Croutons

Sprinkle the chicken breasts with parsley, salt, pepper and a dash of garlic powder. Roll, secure and saute till golden brown. Place cooked, rolled chicken breasts in large buttered casserole. Add drained artichoke hearts (break in half). Combine mayonnaise, wine, soup and lemon juice. Pour over chicken and artichokes. Spread cheese evenly over top. Follow with croutons. I use a prepared herbed stuffing mix for my topping. Bake at 350° for 30 minutes. Should be golden brown on top and bubbly. *Serve with buttered spinach noodles.*

CHICKEN & ARTICHOKE CASSEROLE

preparation time: 30 minutes
cooking time: 1 hour 30 minutes
serves: 10

Alison P. Briscoe
Ovation

10 chicken breasts (boneless)
8 Tblsp. butter
2 Tblsp. oil
½ lb. mushrooms, sliced and fresh
1 can cream of chicken soup
1 can cream of mushroom soup
2 Tblsp. flour
2 cups dry white wine
1 cup water

1 cup whipping cream
2 tsp. salt
Fresh black pepper
½ tsp. sage
½ tsp. basil
2 cans artichoke
12 spring onions
4 Tblsp. chopped parsley

Saute chicken breasts in oil and butter until brown about 10 minutes. Remove chicken, place in casserole dish. Saute mushrooms 5 minutes. Stir in flour. Add soup, wine, and water and simmer until it thickens. Stir in cream, salt, sage, basil, and pepper. Pour over chicken. Bake, uncovered in 350° oven (preheated) for 1 hour. Chop spring onions. Mix in artichoke hearts, onion, and parsley. Bake for 30 more minutes. *Serve with white rice, corn-on-the-cob, green salad and Trifle for dessert.*

CHICKEN PIQUANT

preparation time: 15 minutes
cooking time: 1 hour
serves: 4

Kyle Perkins
Saracen

6 split chicken breasts
¼ cup rose wine
¼ cup soy sauce
¼ cup salad oil
2 Tblsp. water

2 Tblsp. brown sugar
1 clove garlic
1 tsp. ground ginger
1 tsp. oregano

Preheat oven to 350°. Blend all ingredients, except chicken breasts, thoroughly in high speed blender. Place chicken breasts in greased casserole and cover with blended sauce. Bake covered for 1 hour.

CREAM CHICKEN CASSEROLE

preparation time: 20 minutes
cooking time: 1 hour and 15 minutes
serves: 6

Pamela McMichael
Pieces of Eight

1 (10 ¾ oz.) can cream of chicken
    soup
1 cup small curd cottage cheese
¾ cup uncooked rice
1 cup whipping cream
½ cup chopped celery
1 (10 oz.) pkg. frozen broccoli
    spears

½ cup chopped onion
1 tsp. basil leaves
1 garlic clove, minced
Dash salt and pepper
2 lbs. chicken breasts, boned and
    skinned

Combine soup, cottage cheese, rice and cream. Mix well. Stir in celery, onion and seasonings. Pour into 11¾ x 7½ inch baking dish. Arrange chicken breasts over rice mixture. Cover. Bake at 350° for 1 hour and 15 minutes. *Serve topped with hot broccoli.*

PARSLEY-SESAME CHICKEN BREASTS

preparation time: 20 minutes
Cooking time: 15-20 minutes
serves 4:

Pamela McMichael
Pieces of Eight

1 egg
1 Tblsp. milk
½ cup fine, dry bread crumbs
¼ cup sesame seeds
1 Tblsp. chopped fresh parsley
1 tsp. salt

⅛ tsp. green pepper
2 whole chicken breasts, skinned,
    boned and halved
2 Tblsp. vegetable oil
2 Tblsp. unsalted butter
Lemon wedges

Whisk egg and milk in small bowl. Combine bread crumbs, sesame seeds, parsley, salt and pepper. Dip each chicken breast into beaten egg mixture, then dip into bread crumb mixture. Refrigerate 10 minutes. Heat oil and butter and brown chicken breasts for 2 minutes on each side. Lower heat to medium and continue to cook till tender, turning breasts once, 4 to 5 minutes per side. Remove to large platter and garnish with lemon wedges. Serve immediately. *Serve with Radish-Olive Salad, Green Beans in Soy Sauce, followed by Broiled Grapefruit and Grapes.*

CHICKEN BREASTS ROMANO

preparation time: 30 minutes   Lori Moreau
cooking time: 45 minutes   Alberta Rose
serves: 6

6 boneless chicken breasts
2 cups tomato juice
1 medium onion, chopped
½ lb. mushrooms, sliced and
  sauteed
2 Tblsp. vinegar
2 Tblsp. sugar

Garlic powder
Oregano
Basil
1 cup grated Romano
2 Tblsp. chopped parlsey
Flour
4 Tblsp. oil

Skin chicken breasts and fold edges under. Dredge in flour and brown with folded edges down (this will seal edges to keep the shape of the chicken round). Then brown smooth side. Flip chicken over after browning to have smooth side up. Pour tomato juice over. Sprinkle with vinegar, sugar, garlic, oregano and basil. Spread onion and sauteed mushrooms over chicken. Sprinkle with 4 Tblsp. of the Romano and top with chopped parsley. Cover and simmer 45 minutes. Remove chicken to serving platter and top with remaining Romano.

CHICKEN BREASTS IN WHITE WINE

preparation time: 20 minutes   Ann Doubilet
cooking time: 25 minutes   Spice of Life
serves: 6

8 chicken breasts
Flour for dredging
Salt
Pepper
⅔ cup olive oil

3 Tblsp. butter
2 tsp. minced garlic
⅔ cup dry white wine
Juice of 2 lemons
3 Tblsp. chopped parsley

Dredge chicken in seasoned flour. Heat oil in pan and cook only as much chicken as fits in one layer. Start skin side down for 10 minutes, turn and cook 12 more minutes. Remove to do next batch. Pour fat from pan, add cooked chicken and butter. Add garlic and wine and bring to a boil. Add lemon juice and parsley. Cover and cook until done (about 5 more minutes).

SWEET & SOUR CHICKEN

preparation time: 25 minutes   Roz Ferneding
cooking time: 40 minutes   Whisker
serves: 6

6 large single chicken breasts   ½ tsp. ground ginger
2 Tblsp. flour   ¼ cup soy sauce
1 tsp. garlic salt   2 Tblsp. catsup
2 Tblsp. oil   1 Tblsp. vinegar
½ cup chicken broth   1 tsp. honey
1 (20 oz.) can pineapple chunks   1 (6 oz.) pkg. frozen Chinese pea
2 tsp. cornstarch   pods
½ tsp. curry powder   2 Tblsp. sliced green onions

Bone chicken. Combine flour and garlic salt. Coat chicken breasts. Cook chicken in hot oil until brown. Add chicken broth; cover and cook slowly for 15 minutes. Drain pineapple. Save syrup. Combine ¼ cup pineapple syrup with cornstarch, curry and ginger. Blend in soy sauce, catsup, vinegar and honey. Pour over chicken and blend. Simmer, uncovered, 20 minutes longer, or until chicken is barely tender. Add drained pineapple and pea pods. Cook 5 minutes longer. Arrange chicken and sauce on serving platter. Sprinkle with green onions.

ELSIE REICHART'S BOO LOO GAI CHICKEN

preparation time: 15 minutes   K.A. Strassel
cooking time: 1 hour   Great Escape
serves: 6

5 lbs. chicken breasts, skinned   ½ cup soy sauce
2 green peppers, sliced   2 Tblsp. sugar
2 cups pineapple chunks with   Garlic salt
   syrup   Pepper
1 Tblsp. cornstarch   Slivered almonds, optional

Preheat overn to 325°. Heat soy sauce and pineapple syrup, thicken with cornstarch. Add sugar, garlic salt and pepper to taste. Put chicken in 9x13 pan, top with green pepper and pineapple chunks and almonds, pour syrup over all. Bake at 325° for one hour. *Serve with rice.*

CHICKEN APRICOT

preparation time: 20 minutes   Patty Dailey
cooking time: 50 minutes   Western Star
serves: 6

6 chicken breasts, boned
¾ cup dried apricots
Golden raisins
¾ cup water
1½ tsp. grated orange peel

¼ cup orange juice
3 Tblsp. dark corn syrup
1 Tblsp. cider vinegar
1 Tblsp. soy sauce
½ tsp. ground ginger

Preheat oven to 400°. Place 2 dried apricot halves and several golden raisins in center of each breast. Fold sides over and fasten with a toothpick. Bake 40 minutes. Sauce: Cook apricots in water on high for 3 to 4 minutes. Turn to low and cook 3 to 5 minutes. Drain. Combine apricots and remaining ingredients in a blender. Blend on low until smooth. Baste chicken breasts as they cook. *Serve over rice.*

CHICKEN VANITY

preparation time: 30 minutes   Jan Robinson
cooking time: 1 hour   Vanity
serves: 6-8

¼ cup sherry
¼ cup raisins
10 boneless chicken breasts
2 tsp. paprika
Pepper, to taste
¼ cup butter
2 Tblsp. oil
2 cloves garlic, minced

2 cans mandarin oranges
1 cup chicken bouillon
2½ cups fresh mushrooms, sliced
3 Tblsp. soy sauce
1 tsp. ground ginger
2 Tblsp. cornstarch
½ cup plain yoghurt

Sprinkle chicken breasts with paprika and pepper. Preheat electric skillet or large frying pan. Meantime soak raisins in sherry. Melt butter and oil in skillet and brown chicken on both sides. Add chicken bouillon, juice from oranges and minced garlic. Cover and simmer 30 minutes. In a small skillet saute mushrooms lightly in a little butter. Remove chicken to preheated serving dish. Blend cornstarch, ginger and soy sauce with a little water, add to skillet and cook until thickened, stirring constantly. Add drained oranges and mushrooms to heat, and gradually stir in yoghurt. Pour sauce over chicken and serve. *Wild rice and peapods make a good accompaniment.*

CHICKEN SUPREME

preparation time: 20 minutes
cooking time: 30-40 minutes
serves: 8

Maureen Stone
Cantamar IV

8 boneless chicken breasts
3 Tblsp. melted butter
2 Tblsp. orange marmalade

Any bread dressing, flavored with
onions and sage

Remove skin from breasts. Pound slightly to make uniform in size. Place about two tablespoons of dressing on each flattened breast. Roll up jelly roll style tucking in ends. Place seam side down on lightly greased baking pan. Brush all over with melted butter and marmalade. Bake at 350° for 30-40 minutes. *Serve with a tossed Green Salad, Deviled Tomatoes and steamed broccoli.*

CHICKEN TERIYAKI

preparation time: 20 minutes
cooking time: 12-15 minutes
serves: 6

Ann Doubilet
Spice of Life

6 whole chicken breasts, boned
with skin left on
2 cups teriyaki sauce
¼ cup teriyaki glaze (recipe
below)
4 tsp. powdered mustard

Fresh parsley
Glaze:

¼ cup teriyaki sauce
1 Tblsp. sugar
2 tsp. cornstarch mixed with 1
Tblsp. cold water

Heat grill — can be charcoal, gas, or regular broiler. Make glaze: combine ¼ cup teriyaki sauce and sugar. Bring almost to boil, reduce heat and stir in diluted cornstarch. Cook, stirring until thickened to clear glaze. Set aside in bowl. Dip chicken breasts in teriyaki sauce. Broil skin side up, 2-3 minutes, dip again and broil other side 2-3 minutes. Dip again and broil until a golden brown, 4-5 minutes. Place breasts on carving board and slice across grain, skin side up, into 2½-inch slices. Arrange on plate keeping sliced breast in its original form. Pour glaze over and garnish with a small spoonful of mustard and parsley on side. *Serve with rice (Japanese, if possible) and thinly sliced, peeled and seeded cucumber, marinated in rice vinegar and sugar.*

CHICKEN MANDARIN

preparation time: 30 minutes   Jessica Adam
cooking time: 1 hour 15 minutes   Nani Ola
serves: 4-6

2 cut-up chickens	**2 tsp. salt**
½ cup butter	**1 tsp. ginger**
1 cup mushrooms, sliced	**2 tsp. dry mustard**
4 Tblsp. flour	**1 can mandarin oranges**
2½ cups orange juice	**2 Tblsp. parsley, chopped**
4 Tblsp. brown sugar	

Brown chicken in butter. Remove to casserole. Add mushrooms to pan and saute 5 minutes. Scatter over chicken. Turn heat to low and whisk in flour and orange juice. Add remaining ingredients except oranges and parsley. Stir until thickened. Pour over chicken and bake at 350° for 45 minutes. *Serve chicken with oranges and parsley on top for garnish.*

EXOTIC GINGER CHICKEN

preparation time: 40 minutes   Barbara Haworth
marinating time: 3 hours   Ann-Marie II
cooking time: 50 minutes
serves: 6

5 lbs. chicken, legs and thighs	**2 tsp salt**
1 cup rum	**1 tsp. pepper**
2 large onions, chunks	**1 cup soy sauce**
4 cloves garlic, crushed	**1 Tblsp. arrowroot or cornstarch**
2 inches fresh sliced ginger	**¼ cup peanut oil**
2 Tblsp. chives, chopped	**1 can chinese ginger pickles**

Cut off ends of chicken legs, chop legs and thighs in half. Marinate chicken in rum for several hours or overnight. Drain chicken, save rum. Brown chicken in oil and set aside. Saute onions and garlic till light brown, 2 minutes. Add ginger, salt, pepper and chives. Blend arrowroot in soy sauce; add rum and onions mixture. Pour mixture over chicken. Bake 30 minutes in 350° oven. Turn chicken several times to coat. 5 minutes before serving, add pickles; heat for 2-3 minutes. Turn oven off and let stand for 10 minutes. *Serve with rice.*

CHICKEN KIEV

preparation time: 1 hour (plus freezing)   Paula Taylor
cooking time: 15 minutes   Viking Maiden
serves: 10-12

12 boned, skinless chicken breasts   **1 cup flour**
1½ cups softened butter   **3 eggs**
2 Tblsp. chopped, dried parsley   **1 Tblsp. water**
2 Tblsp. dried chives   **2 cups plain breadcrumbs**
½ tsp. pepper and salt

In small bowl combine butter (softened), parsley, chives and crushed garlic cloves, pepper and salt, spread on waxed paper and freeze. This should be done early in the day or the day before, a good time to do it might be when you are making garlic butter for another dish. Pound the chicken on the skin side until ¼-inch think, trying not to break the chicken. A smooth mallet or large bottle will do. Cut the firm butter into ¾" x 3" strips, place one strip lengthwise in the center of one breast, roll up and secure with toothpicks. Dip the rolled breasts into the flour and then in egg and flour mix and then into breadcrumbs. Refrigerate for 1 hour or more. Cook in deep hot oil for 10-15 minutes, until firm to touch. This recipe is relatively easy if done in stages, but either way, it is well worth the effort. *Serve with Marinated Artichoke Hearts, Potato Lynaise and Banana Chocolate Cream.*

FOWL'D UP HAM

preparation time: 1½ hours   D.J. Parker
cooking time: 1 hour   Trekker
serves: 4

4 large chicken breasts   **White Sauce:**
4 thin slices ham
4 slices Provolone cheese   **3 Tblsp. oil**
4 slices uncooked bacon   **3 Tblsp. all purpose flour**
Salt, to taste   **Approximately 2 cups cold milk**
Pepper, to taste   **Minced parsley**

Skin, bone, and pound the chicken breasts flat. Place one slice of ham, then cheese atop each chicken breast. Roll each snug and wrap with bacon. Secure with toothpicks and place in shallow baking dish. Bake at 350° for an hour or until tender. WHITE SAUCE: Heat oil and mix in flour. Slowly blend in cold milk. Heat until thick and creamy, stirring constantly. Add salt, pepper, and parsley. Serve hot over meat. Garnish with fresh parsley. *Serve with Festive Baked Tomatoes, steamed fresh broccoli, caesar salad and Creme de Menthe Pears.*

POULET CORDON BLEU

preparation time: 20 minutes   *Kathi Strassel*
cooking time: 20 minutes   *Great Escape*
serves: 6

6 boneless chicken breasts
12 slices prosciutto or thin sliced
   ham
12 slices gruyere or swiss cheese
1 cup flour
Salt and Pepper
Tarragon

Parmesan
1 cup seasoned bread crumbs
3 eggs
3 Tblsp. safflower or vegetable oil
½ cup dry white wine
½ cup water

Skin breasts and cut in half. On each piece of chicken, place a slice of
prosciutto and cheese. Roll up and secure with toothpicks. Roll each
piece in flour seasoned with salt, pepper, tarragon and parmesan. Dip in
beaten eggs, roll in bread crumbs. Heat oil in heavy skillet and saute
chicken over medium heat, turning to brown evenly (about 10 minutes).
Add water and wine, reduce heat, cover and simmer until done (about 10
minutes). Serve as is, or with gravy made from pan drippings. *Serve with
Spinach Salad, rice, and honey and ginger carrots.*
*Gill Case on M/V Polaris, suggests that after the chicken is cooked and
removed, to then melt butter add capers and lemon juice to taste, instead
of the water and wine.*

SEAFOOD

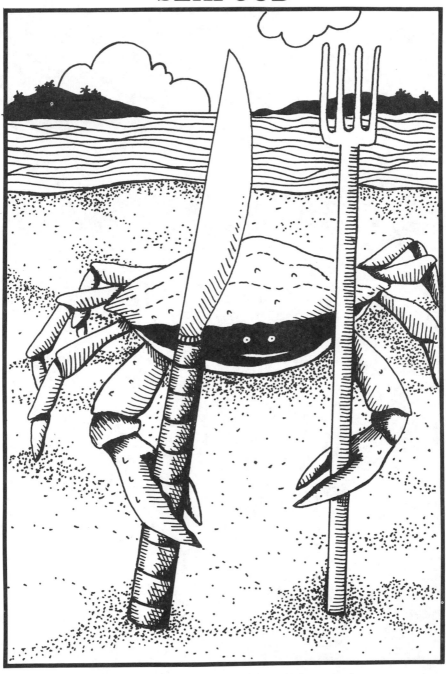

SUGGESTED MENUS

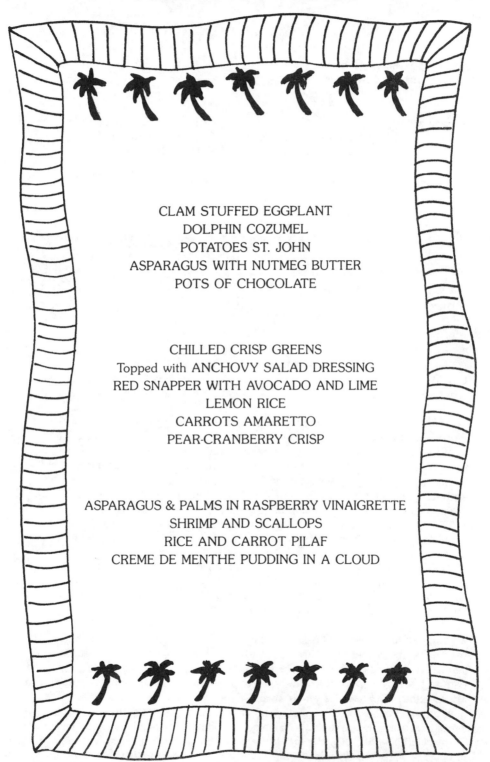

CLAM STUFFED EGGPLANT
DOLPHIN COZUMEL
POTATOES ST. JOHN
ASPARAGUS WITH NUTMEG BUTTER
POTS OF CHOCOLATE

CHILLED CRISP GREENS
Topped with ANCHOVY SALAD DRESSING
RED SNAPPER WITH AVOCADO AND LIME
LEMON RICE
CARROTS AMARETTO
PEAR-CRANBERRY CRISP

ASPARAGUS & PALMS IN RASPBERRY VINAIGRETTE
SHRIMP AND SCALLOPS
RICE AND CARROT PILAF
CREME DE MENTHE PUDDING IN A CLOUD

KING CRAB AND ARTICHOKES

preparation time: 15 minutes
cooking time: 10 minutes
serves: 4-6

Barbara Haworth
Ann-Marie II

2 cans artichokes
2 cans Alaskan King Crab, or fresh
1/2 lb. fresh button mushrooms

1 can mushroom soup
1/2 soup can water
1 Tblsp. white wine

Combine and simmer all ingredients on top of stove in a saucepan over low heat until thoroughly heated. *Serve over rice.*

SOFT SHELL CRABS SAUTEED

preparation time: 20 minutes
cooking time: 10 minutes
serves: 6-8

Jan Robinson
Vanity

8 large soft-shell crabs
1 cup all-purpose flour, seasoned
Clarified butter
Canola oil
Kitchen scissors

Seasoning:
1/4 cup fresh lemon juice
1/4 cup clarified butter
1/2 tsp. freshly ground pepper
1 Tblsp. chopped fresh parsley

You need to clean the crabs as soon as you get them. Begin by rinsing the live crabs thoroughly under cold running water. Rest the crab on your left hand (if you are right handed) with its stomach resting on your palm and its claws dangling over the sides of your hand. With scissors in your other hand snip off the head and eyes. Lift up a side wing of the top shell; cut off the fingerlike gills attached to the body on each side. Leave on the little blobs of stored fat located under the wings. Lift up the apron and snip it off. To force out the sand sack and the tomalley (greenish kind of stuff), squeeze the center of the crab. Rinse the crab again and pat dry. Place the cleaned crabs stomach-side up on a plate so the backs will stay moist.

Note: Crabs are best when they are cooked immediately after cleaning. Although you may clean the crabs ahead of time and refrigerate them. (If you refrigerate them combine two cups milk and 1 Tblsp. dried tarragon and let the crabs soak in the milk for 1 to 2 hours - this will make them moist and plump.)

To Saute the crabs: Heat a large cast iron pan over medium-hot heat. Dredge the soft shell crabs in flour and shake off the excess. When the pan is ready add 1/2 canola oil and 1/2 clarified butter to 1/4 inch deep. Lay crabs in a single layer in the pan, stomach-side up. **Be careful and keep your distance from the pan as the crabs cook, because the crabs sometimes explode!** Brown the crabs lightly, then turn and finish cooking. About 1 minute on each side. Remove the crabs and keep them warm as you saute the remainder. Add more oil and butter, as needed. *To serve: In a small bowl, stir together the lemon juice, clarified butter, and pepper; brush over the crabs and sprinkle with chopped parsley. Serve immediately.*

CARIBBEAN CRAB

preparation time: 15 minutes
cooking time: 40 minutes
serves: 4-6

Cheryl Anne Fowler
Vanity

4 Tblsp. butter
2 Tblsp. olive oil
1 large onion, chopped
1 green pepper, chopped
2 Tblsp. flour
Salt and pepper
1 bouquet garni (celery, parsley, thyme)
1 clove garlic, finely chopped

1 small thin strip lemon peel
Dash Tabasco sauce
¾ lb. flaked crab meat
12 large cooked shrimp, sliced
1 (8 oz.) can okra and juice
1 large can italian peeled tomatoes
1 cup milk
½ cup cream

Combine butter and olive oil in a saucepan and heat until butter has melted. Add chopped onion and green pepper and saute until tender. Blend in flour and cook, stirring continuously, until smooth and thickened. Add salt and pepper, to taste, bouquet garni, garlic, lemon peel and Tabasco sauce. Stir in flaked crab meat, sliced shrimp, okra and tomatoes, and heat to boiling point. Cover saucepan and simmer for 20 to 25 minutes. Gradually stir in milk and cream; cook a few minutes more over medium heat, stirring occasionally, until heated through. Correct seasoning, remove bouquet garni and serve immediately.

SEAFOOD AMOEBA

preparation time: 10 minutes
cooking time: 20 minutes
serves: 6

Fran Bryson
Amoeba

3 lbs. scallops
1 cup juice from scallops
2 Tblsp. flour
1 lb. shrimp
2 green onions, sliced ½-inch pieces

3 cups milk
Salt and pepper
Paprika
2 Tblsp. butter

Saute onions lightly in butter, stir in flour, then add milk and scallop juice, thicken if necessary with cornstarch in double boiler. At the last minute add scallops and shrimp, cook until shrimp are tender but firm. *Serve immediately on pastry shells or with points of toast.*

COQUILLES ST. JACQUES

preparation time: 30 minutes
cooking time: 1 hour
serves: 8

Jeannie Drinkwine
Vanda

2 lbs. sea scallops (save juice)
2 carrots, chopped
1 large onion, chopped
½ lb. fresh mushrooms, sliced
Butter
Juice from 2 lemons

1 cup white wine
Flour
Salt and black pepper, to taste
4 cups of mashed potatoes
3 Tblsp. heavy cream

Shake scallops in flour and saute in butter for approximately 10 minutes (do not overcook). Set aside. Saute carrots and onion in butter then add scallop juice, white wine and ½ of the lemon juice, salt and black pepper. Simmer ½ hour. Drain vegetables and throw away — saving just the liquid. Add cream, mushrooms and balance of lemon juice and simmer additional 15 minutes. Mix in sauteed scallops and put mixture into individual scallop shells or ramekins. Put prepared mashed potatoes into pastry bag fitted with large star tube and pipe a border around the edge of each shell. Put under broiler for a few moments to glaze and brown potatoes and serve immediately.

CONCH CREOLE

preparation time: 30 minutes
cooking time: 30 minutes
serves: 4-6

Harriet Beberman
Avatar

6-8 medium conch
1 clove garlic
2 Tblsp. olive oil or margarine
1 onion, chopped
1 green pepper, chopped
1 red pepper, chopped
2 stalks celery, chopped
1 (3 oz.) can chopped mushrooms

1 can stewed tomatoes, or fresh
    tomatoes
1 (8 oz.) can tomato sauce
¼ cup catsup
Pinch sugar
Dash hot sauce
Salt and pepper, to taste

Pound conch until tender, cut into small bites and set aside. Brown garlic, onions, peppers, and celery in olive oil or margarine until tender. Add can tomatoes, mushrooms, tomato sauce, catsup, sugar, hot sauce, salt and pepper. Heat until bubbly. Turn heat to low and simmer about 15 minutes. Just before serving add conch and heat until conch are cooked through, about 10 minutes. May be made ahead of time and frozen. *Serve over fluffy white rice, green peas, french bread and Simply Sinful Torte.*

SCALLOPS IN SAFFRON CREAM SAUCE

preparation time: 15 minutes   *Terry Boudreau*
cooking time: 20 minutes   *High Barbaree*
serves: 6

2 lbs. sea scallops
1 large shallot, minced
⅛ tsp. powdered saffron
1 Tblsp. butter
2 large tomatoes, peeled, seeded
  and chopped

½ lb. mushrooms, thinly sliced
2 Tblsp. brandy
2 Tblsp. dry white vermouth
1 cup heavy cream
6 Tblsp. butter

Rinse 2 lbs. sea scallops and pat dry. In a flameproof casserole or large fry pan cook large shallot with ⅛ tsp. powdered saffron in 1 Tblsp. butter over low heat until the shallot is softened. Add the scallops, tomatoes and mushrooms and 2 Tblsp. each of brandy and white vermouth and cook over very low heat, covered for ten minutes. Transfer scallops with slotted spoon to a heated serving dish and keep them warm, covered. Add 1 cup of heavy cream to the pan, stirring and reduce the sauce over high heat to ¾ cup. Swirl in 6 Tblsp. butter cut into bits, season the sauce with salt and pepper and pour over scallops. Garnish with sliced lemon and parsley.

Serve with Broccoli with Mushrooms, Clam Stuffed Eggplant as an Appetizer and Quick Chocolate Mousse for dessert.

SEA SCALLOPS PROVENCAL

preparation time: 20 minutes   *Jeannie Drinkwine*
cooking time: 20 minutes   *Vanda*
serves: 6

2 lbs. sea scallops, halved if large
4 Tblsp. olive oil
2 cloves garlic, finely chopped
4 Tblsp. shallots, chopped
2 (16 oz.) plum tomatoes, well
  drained and chopped

½ tsp. salt
¼ tsp. freshly ground black pepper
½ tsp. thyme
1 tsp. basil
2 Tblsp. parsley, chopped

Heat the olive oil in a skillet. Add the garlic and shallots and cook until tender but NOT browned. Add the tomatoes, salt, pepper, thyme and basil. Simmer uncovered about 10 minutes. Add the parsley and scallops and cook, stirring occasionally, 3 to 5 minutes or until scallops are opaque. Do not overcook. *Serve with rice.*

SCALLOPS IN WINE

preparation time: 15 minutes　　　　　　　　*Jessica Adam*
cooking time: 25 minutes　　　　　　　　　　*Nani Ola*
serves: 4-6

2 lbs. scallops	2 Tblsp. garlic, minced
½ cup flour	2 tsp. salt
1 cup butter	1 tsp. white pepper
1 small white onion, minced	2 Tblsp. lemon juice
1 lb. mushrooms, sliced	1½ cup white wine

Shake scallops in flour. Lightly brown in ½ cup butter. Melt ½ cup butter in clean pan. Saute onion for 5 minutes. Add mushrooms and garlic. Season with salt and pepper. Continue to saute for 5 minutes more. Add scallops, lemon juice and wine. Simmer for a few minutes until heated through. *Serve with Lentil-Vegetable Soup, Broccoli in Lemon Sauce and rice.*

SCALLOPS AU PROVENCE

preparation time: 20 minutes　　　　　　*Pamela McMichael*
cooking time: 10-12 minutes　　　　　　　*Pieces of Eight*
serves: 4

12 large sea scallops (about 1½ lbs.)	¼ tsp. sugar (optional)
	2 Tblsp. apple cider of juice
6 to 8 slices bacon (halved crosswise)	1 Tblsp. lemon juice
	4 large sprigs watercress (optional)
Salt	
Pepper	
2 large apples (Granny Smiths, Jonathans, Cortlands or Winesap)	

Heat oven to 425°. Blanch bacon strips by dropping into boiling water and cooking 1 minute. Drain. Place scallops in single layer in buttered baking dish. Season with salt and pepper. Place bacon on top. Bake till scallops are cooked through and bacon is crisp, about 10 to 12 minutes for large scallops. While scallops are baking cut apple into 8 wedges. Pare and remove core. Sprinkle apple with lemon juice. Heat butter in skillet, add apple wedges and sprinkle with sugar (if using). Saute apple wedges till lightly browned, about 5 minutes. Arrange scallops and apple wedges on plates and garnish with watercress. *Serve with Steamed Vegetables in Malt Vinegar Dressing, and Light Chocolate Pudding.*

SAUTÉED SCALLOPS WITH WINE

preparation time: 10 minutes
cooking time: 15 minutes
serves: 6

Ann Doubilet
Spice of Life

2 lbs. scallops, fresh or frozen
4 Tblsp. butter
Flour for dredging

⅔ cup dry white wine
Parsley, chopped

Wash scallops and lay out on paper towels to dry. Very lightly toss with flour a few at a time. Heat butter and cook in small batches, taking up only ½ of pan so scallops brown lightly. Remove, keep warm, and continue lightly dredging and lightly browning. Deglaze pan with dry white wine, add parsley and pour over scallops. *Serve with rice.*

LOBSTER THERMIDOR

preparation time: 30 minutes
cooking time: 15 minutes
serves: 4

Cheryl Anne Fowler
Vanity

2 (2 lb.) lobster, cooked
½ cup butter
2 small onions, finely chopped
½ cup all-purpose flour
1¼ cups milk
½ cup grated Cheddar cheese

2½ Tblsp. dry white wine
¼ tsp. paprika
Salt and pepper
Grated parmesan cheese
Lettuce, watercress, lemon slices
**   for garnish**

Cut boiled lobsters in half carefully and remove meat. Discard intestines. Twist off claws and remove meat. Clean shells thoroughly and rub with oil to make them shiny. Cut meat into pieces ½ inch long. Heat half the butter in a skillet and saute meat. Meanwhile, in another pan heat remaining butter and saute onion till soft. Stir in the flour and cook over low heat for about 30 seconds. Remove from heat and stir in milk. Cook over medium heat, stirring constantly, to a boil and gently simmer for 2 minutes. Remove from heat and stir in cheese, wine and paprika. Season. Cook over low heat, stirring, for about 1 minute. Pour the cheese sauce over the hot lobster in the skillet. Cook over low heat for a few minutes. Remove from heat. Place the cleaned lobster shells on a broiler rack and fill them with the lobster mixture. Sprinkle heavily with grated Parmesan. Broil in a pre-heated broiling compartment, about 4 inches from broiler until sauce is bubbling and top is golden brown. *Serve immediately garnished with lettuce, watercress, and lemon slices.*

"HULLO" LOBSTER

preparation time: 30 minutes   Guy Howe
cooking time: 6-8 minutes   Goodbye Charlie
serves: 4

1 lobster (1½-2 lbs.)   2 Tblsp. cornstarch
¼ lb. lean pork, ground   2 Tblsp. sherry
1 scallion, minced   2 Tblsp. soy sauce
2 sliced fresh ginger root, minced   ½ tsp. salt
2 eggs   ½ cup water

With a cleaver, chop lobster lengthwise in half; clean. Chop each half, shell and all, in 1½ inch sections. Combine pork, scallion and ginger root. Beat eggs lightly. Then combine cornstarch, sherry, soy sauce, salt and water, blending well. Heat oil. Add lobster sections and stir-fry over medium heat until the shell turns bright red (about 3 minutes). Add pork mixture and cook, stirring, over low heat until meat is cooked through (3 to 5 minutes). Serve immediately.

ADMIRALS STIR FRIED LOBSTER

preparation time: 15 minutes   Jan Robinson
cooking time: 8-12 minutes   Vanity
serves: 4

1 lb. lobster meat   2 Tblsp. soy sauce
1 Tblsp. cornstarch   1 Tblsp. sherry
¼ cup water   2 Tblsp. oil
¼ lb. fresh mushrooms   3-4 Tblsp. oil
2 scallion stalks   ½ tsp. salt
1 garlic clove

Cut lobster meat in 1 inch cubes. Blend cornstarch and cold water to a paste; add to lobster cubes and toss gently to coat. Slice mushrooms. Cut scallion stalks in 1 inch sections. Mince garlic; then combine with soy sauce and sherry. Heat 2 Tblsp. oil. Add scallions; stir fry a few times. Add mushrooms and stir-fry until nearly done (2-3 min.). Remove vegetables from. Heat 3-4 Tblsp. oil, add salt and lobster cubes. Stir-fry until cubes begin to curl at the edges (3-4 min.). Quickly stir in garlic-soy mixture. Then return mushrooms and scallions. Stir-fry to reheat and blend flavors (2 min. or more). Serve at once.

LOBSTER MONIQUE

preparation time: 15 minutes
cooking time: 20-25 minutes
serves: 4

Erica Benjamin
Caribe Monique

2 finely chopped pimentos
1 tsp. salt
½ tsp. worcestershire sauce
2 Tblsp. dry wine
¼ tsp. prepared mustard
1 lb. boiled lobster meat

6 Tblsp. butter
6 Tblsp. flour
2 cups hot milk
2 Tblsp. cracker or bread crumbs
2 Tblsp. butter

Mix together pimentos, salt, Worcestershire, wine and mustard. Cut lobster meat into ½ inch chunks. In a saucepan, melt 6 Tblsp. butter. Add flour and stir constantly for a few seconds. Gradually stir in milk and continue stirring over moderate heat. Add Worcestershire mixture and lobster meat. Stir constantly until mixture boils and thickens. Pour mixture into a glass baking dish, cover lightly with crumbs and dot with 2 Tblsp. butter. Bake at 400°F until crumbs are browned.

LANGOSTINOS AL JEREZ
(Shrimp in Sherry)

preparation time: 15-20 minutes
cooking time: 15 minutes
serves: 4-6

Silvia Kahn
Antipodes

2 lbs. jumbo shrimp, cooked and
    peeled
¼ cup olive oil
¼ cup butter
3 onions, sliced
Fresh garlic to taste
1 box frozen baby peas
1 can red spanish peppers
    (pimientos)

1 cup sour cream
1 Tblsp. corn flour
Salt and pepper
Saffron, to taste
1 cup Dry Sack Sherry
Fresh parsley

In large frying pan heat oil and butter. Saute onions, garlic, peas and chopped red peppers. Mix a small amount of water with corn flour and combine with sour cream. Add to sauteed vegetables. Stir. Add saffron, salt, pepper and sherry. Stir and add cooked shrimp until heated through. Ready to serve with rice. Sprinkle with fresh chopped parsley. Serve with Melon con Hamon, Arroz Amarillo and Quesillo Flan Caramelo, Espanola Vino and Flamenco Musik.

SHRIMP IN A CURRY SAUCE

preparation time: 20 minutes
cooking time: 20 minutes
serves: 4

Jan Robinson
Vanity

1 lb. shrimp, large
2 medium onions, chopped
2 Tblsp. olive oil
1 Tblsp. butter
2 clove garlic, minced
2 tsp. curry powder
2 Tblsp. flour
¼ cup white wine
1½ cups fish or chicken stock

1 Tblsp. tomato puree
2 Tblsp. Mango chutney or apricot
    jam
Juice of 1 lemon
4 Tblsp. butter
⅓ cup heavy cream
3 cups cooked rice, hot
2 (17 oz.) cans Lesueuer Early
    Peas, heated

Saute onion and garlic in oil and butter. Stir in curry powder. Blend in flour and cook for a few minutes. Gradually add stock and wine and stir until sauce thickens. Add tomato puree, chutney and lemon juice. Simmer for 10 minutes or so, while preparing shrimp. Peel, rinse in cold water and pat dry. Heat the butter in a skillet and saute the shrimp. Blend in the curry sauce and bring to a boil. Stir in the cream and remove skillet from the heat. Serve over rice and with peas, or individual plates. Take 2 6oz. cups (I use pewter Jefferson cups) and fill each with ice. Discard ice from one cup and pack in enough rice for one serving. Tip rice on to serving plate. (Cold cups prevent rice sticking) Repeat process until you have a nice firm mound of rice on each plate. Spoon curry sauce over rice (gently). Place shrimp around base and one on top. Place green peas between shrimp. *Fruit Salad with Lime Sauce makes a great appetizer and for dessert, New Zealand Pavlova.*

SHRIMP CONGA

preparation time: 15 minutes
cooking time: 15-20 minutes
serves: 4

Roz Ferneding
Whisker

1 lb. shrimp
Lemon or lime
½ cup butter, melted

3 oz. cream cheese
1 oz. Roquefort or Blue cheese

Peel and butterfly shrimp. Place shrimp in one layer in baking pan; squeeze lemon or lime juice over shrimp. Mix butter, cream cheese and Roquefort and spread over shrimp. Bake, covered, 15-20 minutes. *This is a stolen recipe!!*

SZECHUAN SHRIMP

preparation time: 30 minutes   *Saracen*
cooking time: 10 minutes
serves: 4

½ cup minced scallions
½ cup minced bamboo shoots
¼ tsp. minced ginger root
3 large garlic cloves, minced
¼ tsp. tabasco
2 Tblsp. sugar
½ cup ketchup
3 Tblsp. dry sherry

1 Tblsp. soy sauce
1½ tsp. sesame oil or 3 Tblsp.
   toasted sesame seeds
1 Tblsp. cornstarch
3 Tblsp. water
1½ cups oil
1½ lbs. shrimp, cleaned and
   deveined

First bowl — scallions, bamboo shoots, ginger, garlic, tabasco
Second bowl — sugar, ketchup, sherry, soy sauce and sesame oil or
seeds
Third bowl — cornstarch and water
Cook shrimp in Wok with oil and drain in strainer, keeping oil. Heat 2
Tblsp. oil and cook first bowl and fry 1 minute. Add drained shrimp and
cook 30 seconds. Pour in ketchup mixture and stir 30 seconds. Add cornstarch and water mixture and stir. *Serve with rice and pea pods or egg
rolls, rumaki, etc. for appetizers.*

SHRIMP AND MUSHROOMS

preparation time: 20 minutes   *Barbara Haworth*
cooking time: 10 minutes   *Ann-Marie II*
serves: 4

¼ cup peanut oil
1½ cups celery, sliced diagonally
   in ½-inch strips
1 lb. fresh mushrooms, sliced
¾ cup scallions, sliced
2 Tblsp. soy sauce
½ tsp. ginger, ground
¼ cup white wine or sherry

¼ tsp. pepper, ground
2 Tblsp. cornstarch
¼ cup cold water
1 cup chicken broth
1¼ lbs. shrimp, peeled and
   deveined
½ pkg. frozen pea pods

Preheat wok or large skillet. Stir-fry celery 3 minutes. Push to the sides or
remove. Stir-fry mushrooms 2 minutes (if canned- stir-fry 1 minute). Push
to the sides. Stir-fry shrimp 3-4 minutes. Dissolve cornstarch in water.
Add seasonings to soy sauce. Add all liquids to vegetables and shrimp.
Stir until thickened. Cover and heat 2 minutes till hot. *Serve with rice.*

GINGER SHRIMP

preparation time: 30 minutes
cooking time: 15 minutes
serves: 6-8

Gill Case
M/V Polaris

½ lb. snow peas
2½ lbs. medium shrimp
6 Tblsp. olive oil
2 Tblsp. dry sherry
1 tsp. salt
Pinch pepper
2 garlic cloves, chopped
2 slices gingerroot, chopped

2 scallions, chopped
½ cup ketchup
4 tsp. sugar
2 Tblsp. Worcestershire
Dash of Tabasco
2 tsp. cornstarch
2 Tblsp. water

Shell and devein shrimp. Heat oil in a large skillet. Add shrimp, sherry, salt and pepper. Stir-fry till shrimp are pink. Remove from pan and keep warm. Add garlic, ginger and scallion; stir-fry a few seconds. Add ketchup, Worcestershire, Tabasco and sugar; stir-fry 1 minute. Stir in cornstarch dissolved in 2 Tblsp. water and simmer stirring till thickened. Adjust seasonings. Return shrimp to pan till heated through. Snow peas can be added to sauce at end of cooking to add color and texture. *Serve with Won Ton Bows, Stir Fried Rice, Stir Fried Mixed Vegetables and Honeydew Melon with your favorite liqueur.*

SHRIMP DIABLO

preparation time: 1 hour
cooking time: 30 minutes
serves: 8

Jeannie Drinkwine
Vanda

1 cup olive oil
2 large onions, diced
4 cloves garlic, minced
4 green peppers, diced
6 Tblsp. flour
4 lbs. cleaned, cooked shrimp

1 (16 oz.) can tomato sauce
2 tsp. salt
½ tsp. hot pepper sauce (or more)
2 tsp. thyme
2 cups dry white wine

Heat oil in Dutch oven. Add onion, garlic and green peppers and cook until onion is golden. Stir in flour and cook slowly, stirring constantly, until flour browns lightly, about 5 minutes. Add shrimp and stir to coat well with roix. Stir in tomato sauce. Add salt, pepper sauce and thyme, then stir in 2 cups wine. Cover and heat through. If sauce is too thick, stir in ¼ cup more wine. Serve over fluffy hot rice.

SCAMPI

preparation time: 20 minutes
cooking time: 15 minutes
serves: 6

Shirley Fortune
Maranatha

48 jumbo shrimp (under 15 count per pound)
6 cloves garlic, crushed
2 tsp. salt
¾ cup butter
¾ cup olive oil
¼ cup minced parsley
2 Tblsp. lemon juice
Freshly ground black pepper

Garnish:
3 lemons, scalloped cut
1 large green pepper, sliced lengthwise (24 slices)
6 tiny parsley sprigs for center of lemons

Use a knife and devein shrimp and butterfly cut through shell back lengthwise, almost to the tail. Leave feet on and shell intact. Melt butter and add all other ingredients, add shrimp when butter is hot (need 2 frying pans to do this amount of shrimp). Cook shrimp until they "butterfly" nicely and are pink — don't overcook. To serve: place ½ cup Lemon Rice (see index) in center of plate and flatten in a circle. Place lemon in center and 4 pepper strips outwards from lemon, giving you 4 sections. Stand 2 shrimp per section, cut flesh side down and tails in the air. Spectacular! *Serve with Lemon Rice for Seafood, Cauliflower, Avocado Salad and Frangelico Velvet.*

SHRIMP SCAMPI

preparation time: 20 minutes
cooking time: 5-8 minutes
serves: 8

Debbie Olsen
Kingsport

2½ lbs. shelled jumbo shrimp
4 garlic cloves
½ cup butter
1 Tblsp. lemon juice

1 tsp. salt
⅛ tsp. pepper
½ cup oregano
¼ cup parsley

In saucepan combine all ingredients except shrimp, heat through. Arrange shrimp in baking dish. Pour butter mixture over and broil 5-8 minutes. Sprinkle with parsley and garnish with lemon slices.

QUICK SCAMPI

preparation time: 10 minutes
cooking time: 15-20 minutes
serves: 6

S.L. Lehman
Lady Sam

30 large or 24 jumbo shrimp (fresh or frozen)
4 cloves garlic
1 large cube butter
2½ Tblsp. cornstarch

1½-2 cups chicken broth (as desired)
1 Tblsp. soy sauce (if desired)
3 Tblsp. Old Tyme Barbadian Hot Pepper Sauce
1 Tblsp. parsley flakes

Shell all shrimp, leave tail if desired, do not pre-boil. Chop garlic finely. Add butter to a wok, electric skillet or your favorite hot cooking skillet. When hot and bubbly, add garlic and quick stir (about 1½ minutes). Add soy sauce and hot pepper sauce, continue to stir another minute. Pour in shrimp and quick stir about 3 minutes. Cover and simmer about 3-5 minutes. In an old jar put chicken broth and cornstarch, shake well; then pour into fixings. Continue to stir until sauce starts to thicken (not too much). Top with parsley flakes. Place wok or skillet directly at table center while hot and serve.

SHRIMP ALBERTA ROSE

preparation time: 30-40 minutes
cooking time: 30 minutes
serves: 6

Lori Moreau
Alberta Rose

2 lbs. jumbo shrimp, peeled and deveined
3 (14 oz.) cans artichoke bottoms
4 oz. butter
¾ lb. mushrooms, sliced
¾ cup grated romano

3 Tblsp. flour
3 Tblsp. butter
1½ cups cream
½ cup white wine
Paprika

Arrange artichoke bottoms (cut side up) in bottom of a 9x13" pan. Sauté mushrooms in butter and spread over artichokes. Sauté shrimp for 3 minutes. Arrange attractively on top of mushrooms. Prepare a white sauce with flour, 3 Tblsp. butter and cream. Add white wine to sauce and pour evenly over shrimp. Sprinkle with romano and paprika. Cover and bake at 350° for 30 minutes. *Serve with Cucumber Soup, rice pilaf, steamed broccoli and Apple Kuchen.*

SHRIMP AND SCALLOPS

preparation time: 20 minutes
cooking time: 20-30 minutes
serves: 8

Gay Thompson
Satori

16 jumbo shrimp
1-½ lbs. scallops
½ cup butter
½ lb. mushrooms
6 green onions
1 clove garlic, pressed

1 lg. bottle capers
½ green pepper
¼ cup lemon juice
1 lemon
parsley

Wash and clean shrimp and scallops. Cut scallops into 1 inch pieces, if large. Slice mushrooms, onions, and lemon. Dice green pepper. Melt butter in a 9 inch roasting pan (deep dish). Add shrimp, scallops, mushrooms, onions, green pepper, lemon juice and pressed garlic; top with drained capers. Butter should surround shell fish so add more if necessary. Bake at 350° 20-30 minutes. Turn fish after 15 minutes to cook evenly. When shimp are pink it is done. Top with sliced lemon and parsley. *Serve immediately in bowls so you can enjoy all that great butter!*

POLYNESIAN BUTTERED SHRIMP & PINEAPPLE

preparation time: 35 minutes
cooking time: 45 minutes
serves: 4

Kandy Popkes
Wind Song Oregon

4 Tblsp. butter
1 lrg. green pepper — seeded &
   sliced in ¼" strips
1-½ lb. uncooked shrimp —
   shelled

1 can pineapple slices
¾ tsp. salt
2 tsp. chili powder
Lime or lemon wedges

Melt butter over medium heat. Add peppers, shrimp and pineapple. Season with salt & pepper & chili powder. Saute just until shrimp turns pink & pepper is tender crisp — about 10 min. *Serve immediately with lime or lemon wedges on hot steamed rice.*

SHRIMP CREOLE BRYSON STYLE

preparation time: 20 minutes   Fran Bryson
cooking time: 30 minutes   Amoeba
serves: 6

3 to 5 lbs. shrimp
1 onion, chopped
1 garlic clove
1 green pepper
2 stalks celery
1 Tblsp. butter
1½ cup tomatoes

1 cup tomato sauce
1 tsp. thyme
1 bay leaf
1 Tblsp. parsley
½ Tblsp. sugar
Salt and pepper

Boil shrimp until tender. Chop and saute onion, garlic, green pepper and celery in butter for 5 minutes. Add remaining ingredients and simmer for 20 minutes. Add shrimp to heat through, about 10 minutes. *Serve over rice.*

CEVICHE

preparation time: 20 minutes   Terry Boudreau
marinating time: 3-4 hours   High Barbaree

3 lbs. fillets of fish (snapper or any
    white meat fish)
3 cups orange juice
1½ cups lemon or lime juice
3 medium onions chopped

3 green peppers chopped
2 tsp. salt
pepper to taste
2 cups ketchup
4 Tblsp. olive oil

Cut fish fillets into small pieces and add all the other ingredients in a ceramic bowl. Mix well. Cover and refrigerate for 3-4 hours. Stir occasionally. *Serve very cold with popcorn.*

SEAFOOD CREPES

preparation time: 20 minutes
cooking time: 25-30 minutes
serves: 4-6

Roz Ferneding
Whisker

6 medium mushrooms sliced
3 Tblsp. green onion, chopped
3 Tblsp. butter
3½ cups cooked seafood
3 (3 oz.) pkgs. cream cheese

1 cup half and half
1 cup shredded swiss cheese
¼ cup chopped green onion
8 cooked crepes (see Index)

Sauté mushrooms and 3 Tblsp. green onion in butter. Add seafood, cream cheese, half and half. Mix and spoon into 8 crepes. Put swiss cheese on top and bake for 20 minutes at 350°. Garnish with green onions.

SEAFOOD CASSEROLE

preparation time: 35 minutes
cooking time: 40 minutes
serves: 4-6

Kyle Perkins
Saracen

1 head broccoli, cooked, broken in
    flowerettes
2-3 lbs. Seafood (lobster, fish,
    shrimp, crab, combination of
    anything is good, leftovers
    fine).
½ lb. Fresh mushrooms, sauteed.

Layer in casserole dish in above order. Pour sauce over, top with buttered corn flakes or bread crumbs and paprika. Bake uncovered 30-40 minutes 375°.

SAUCE:
3 Tblsp. Butter
2-2½ cups milk (room
    temperature)

3 Tblsp. Flour
20 oz. sharp cheddar cheese
Sherry and seasonings to taste.

Melt butter, add flour, gradually add milk and whisk. Add cheese, sherry, seasonings.

GABRIELLES TURBOT

preparation time: 30 minutes　　　　　　　　　　*Gay Thompson*
cooking time: 20-30 minutes　　　　　　　　　　　　　　*Satori*
serves: 8

2½ lbs. turbot (thawed if frozen &
　pressed to rid fish of excess
　water)
1 pkg. Filo dough
½-¾ cup butter
8 ozs. cream cheese, softened
4 ozs. Muenster cheese, grated

½ lb. mushrooms, washed and
　sliced
1 clove garlic, pressed
4 green onions, sliced
4 Tblsp. plus 1 tsp. coarse ground
　pepper
1 pkg. sliced almonds
Parsley, chopped

Butter a 9x13x2 inch oven-proof pan. In small skillet saute mushrooms, onions, garlic and 1 Tblsp. pepper in 2 Tblsp. butter till limp. Have cheese ready and fish pressed before you open the filo — you have to work fast or it dries out. Place a sheet of filo on bottom of buttered pan, letting edges hang over sides. Brush with melted butter. Place another sheet of dough over the first layer at right angles, brush with butter. Continue to place sheets around bottom at different angles, overlapping sides and buttering between layers until you have about 6-10 layers. Place fish in pan and spread with cream cheese, grated muenster and sauteed vegetables. Quickly bring overlapping edges over top of fish. Cover with 2 more filo sheets, buttering between sheets and making sure there is no dough hanging over the outside of the pan. Brown almonds in butter, add tsp. pepper and parsley. Stir and pour over top of dough. Bake at 350° for 20-30 minutes or until filo dough is golden brown. *Serve with Schooner Spinach, a green salad and bananas flambé.*

LAZY SOLE

preparation time: 10 minutes　　　　　　　　　　*Roz Ferneding*
cooking time: 25 minutes　　　　　　　　　　　　　　*Whisker*
serves: 6

3 lbs. Sole
1-2 lemons
6 Tblsp. butter

Italian seasoning
Flour
Toasted almonds — garnish

For this recipe I use Carnation Atlantic Sole, (it's just as good as the expensive stuff). Layer Sole in baking dish, I use the one I'm going to serve it in. Squeeze lemon juice over fish and chunks of butter. Sprinkle a little Italian seasoning over. Bake at 350° for 25 minutes. Drain off all liquid into a pan and thicken with a little flour. Pour over fish and sprinkle with toasted almonds.

CARIBBEAN BUNDLES

preparation time: 25 minutes   *Cheryl Anne Fowler*
cooking time: 20-25 minutes   *Vanity*
serves: 4

4 fresh or frozen sole, flounder or   ¼ cup chopped onion
   other fish fillets, ¼ inch thick   ¼ cup dry white wine
¾ lbs. fresh asparagus or one 8   1 clove garlic, minced
   oz. package frozen asparagus   2 tsp. snipped fresh mint of ½ tsp.
   spears   dried mint, crushed
2 medium tomatoes, peeled and   ½ tsp dried basil, crushed
   cut up (1 cup)   ¼ tsp. salt
½ cup sliced fresh mushrooms
¼ cup thinly sliced celery

Thaw fish, if frozen. Cut the fresh asparagus spears into about 6-inch lengths. In a covered saucepan cook the cut fresh asparagus in a small amount of boiling salted water for 8 to 10 minutes or till almost tender. (or, cook the frozen asparagus spears according to the package directions.) Drain the cooked asparagus well. Sprinkle the fish fillets with a little salt. Place the cooked asparagus across the fish fillets; roll fish fillets around the asparagus and fasten securely with wooden picks. Place the fish rolls, seam sides down, in a 10-inch skillet. To the fish rolls in the skillet, add the cut tomatoes, sliced mushrooms, sliced celery, chopped onion, dry white wine, garlic, fresh or dried mint, basil, and ¼ tsp. salt. Cover the skillet tightly; simmer over low heat for 7 to 8 minutes or till the fish flakes easily when tested with a fork. Remove the fish rolls to a warm platter; remove the wooden picks. Cover the fish rolls and keep warm. In the skillet boil the tomato mixture gently, uncovered, about 3 minutes or till slightly thickened. Spoon the tomato mixture over the fish rolls. Garnish with fresh mint leaves. *This recipe may sound involved, but it's quite simple and looks and tastes terrific.*

FISH CALYPSO

preparation time: 20 minutes   *Kandy Popkes*
cooking time: 15-20 minutes   *Wind Song Oregon*

Turbot, Kingfish, or other firm   Italian Seasoned breadcrumbs
   white fish   Bananas
Butter   Egg
Slivered almonds

Add almonds to breadcrumbs. Dip fish in egg, then roll in breadcrumbs. Fry in butter until golden. Keep warm in oven. Slice bananas in circles and saute in butter until soft. *Serve over the top of fish.*

RED SNAPPER ROLLS

preparation time: 1 hour
cooking time: 35 minutes
serves: 4-6

Shirley Fortune
Maranatha

4 Tblsp. butter
¼ cup chopped onion
1 (3 oz.) jar sliced mushrooms
  (save juice)
1 (7½ oz.) can crab, drained
½ cup crushed saltines
2 Tblsp. parsley, chopped
¼ tsp. pepper
6 red snapper fillets
  (approximately 2 lb.)

Sauce:

3 Tblsp. butter
3 Tblsp. flour
1¼ cup milk
⅓ cup dry white wine
4 oz. grated swiss cheese
½ tsp. paprika
Mushroom liquid

Cook onions in butter until tender, add mushrooms, crab, saltines, parsley and pepper. Spread mix on fish fillets and roll up. Secure with picks. Place seam side down in a pretty baker. Sauce: Melt butter, add flour. Add milk and mushroom liquid to make 1½ cups. Add wine to butter-flour mixture, cook and stir until thick. Pour over fillets and bake 25 minutes at 400°. Then sprinkle with cheese and paprika and bake 10 minutes more. *To serve, garnish with parsley sprigs between fillets and serve with warm garlic bread, artichokes, semi-dry white wine, and Creme de Menthe Pudding in a Cloud.*

POISSON VERONIQUE

preparation time: 20 minutes
cooking time: 25 minutes
serves: 6

Kathi Strassel
Great Escape

3-3½ lbs. grouper fillets
¾ cup dry white wine
1 cup water
2 Tblsp. chopped scallions
Seasoned sea salt
1½ cups green grapes
2 Tblsp. butter

½ cup cream
2 Tblsp. flour
Garnish:

Grapes
Lemon Wedges

Wash, dry and season fillets. In heavy skillet bring water, wine and scallions to boil. Add fillets, cook 3-4 minutes each side. Remove and drain fillets. Add grapes to boiling stock, poach for 3 minutes, remove and drain. Boil stock to one cup — reserve. Melt butter, add flour to make thick paste. Add cream and reserved stock, bring to boil stirring constantly. Place fillets in ovenproof casserole, cover with sauce. Broil until sauce glazes. Garnish with grapes and lemon wedges.

SEA SWEET SECRET

preparation time: 20 minutes   Sandy Culbert
cooking time: 6-10 minutes   Mystique

**Fresh white meated fish (Spanish
    Mackeral, Grouper, Snapper,
    Scallops are fine as well, or
    Shrimp)
Garlic
Fresh ginger
Tamari or Kikoman sauce, to taste
1 oz. lemon juice**

**½ cup honey, or more
Dash pepper
Garnish:
Lettuce
Watercress
Lemon slices**

Place fish in suitable pan to put under the broiler. Squash the garlic and
ginger together, mix with the lemon, Tamari, honey, and pepper. Spread
over fish and broil. Don't overcook! *Serve this with rice around it and the
juice poured over it (which you can expand by adding a little water or
white wine to the pan juices as in any other gravy). A frame of finely
chopped lettuce, watercress, and lemon makes a lovely presentation.*

FISH FILLETS MEUNIERE

preparation time: 15 minutes   Shari Stump
cooking time: 10 minutes   Eyola
serves: 6

**3 lbs. fish fillets
¼ cup all-purpose flour
¼ tsp. salt
Pinch of white pepper
Juice of 1 lemon**

**1 Tblsp. chopped parsley
2 tsp. chopped fresh chives and/
    or tarragon
Pinch each of salt and white
    pepper
7 Tblsp. butter**

Wash fillets; pat dry with paper towels. Mix flour with salt and pepper.
Coat fillets with flour mixture and pat off excess. Combine lemon juice,
parsley, other herbs, salt and pepper. In a large skillet heat 3 Tblsp. butter until foaming; add fillets. Cook over medium heat 1 to 3 minutes until
golden brown. Brown on all sides. Do not overcook or fish will fall apart.
Remove from skillet and keep warm on hot plate. Add remaining butter;
heat until nut brown. Immediately add lemon juice mixture; swirling pan
quickly to blend with butter. Pour foaming mixture over fish. Serve immediately. Decorate with lemon, orange slices and parsley. Delicious!

RED SNAPPER WITH AVOCADO AND LIME

preparation time: 10 minutes   Jean Thayer
cooking time: 25-30 minutes   Paradise
serves: 6

6 red snapper fillets
Juice of 2 limes
¼ lb. butter
Dash of garlic salt
1 tsp. salt

2 Tblsp. sherry
1 large ripe avocado
1 lime
2 egg yolks

Preheat oven to 350°. Place fillets in glass baking pan. In small saucepan, melt butter, add juice, salts and sherry. Brush fish with sauce and place in oven. Continue to baste 2 or 3 times during cooking time. Cook till fish flakes easily with a fork. Slice peeled avocado and place in small baking dish. Squeeze juice of 1 lime over avocado and bake during the last 5 minutes. Thicken remaining basting sauce with 2 egg yolks, whisking sauce continuiously over low heat. To serve, ladle sauce over fish and garnish with avocado. *Serve with rice, steamed carrots, coleslaw and Pots of Chocolate.*

SALMON/GREEN BEAN SUPREME

preparation time: 20 minutes   Cheryl Anne Fowler
cooking time: 20 minutes   Vanity
serves: 4

4 frozen salmon or other fish
    steaks, cut ¾ inch thick
2 cups water
2 Tblsp. lemon juice
1 tsp. salt
1 (9 oz.) package frozen French-
    style green beans
1 cup skim milk

½ cup process Swiss cheese cut
    into cubes (2 oz.)
2 Tblsp. all-purpose flour
1 Tblsp. butter or margarine
¼ tsp. salt
1 Tblsp. dry sherry
2 Tblsp. slivered almonds

Place frozen fish steaks in large skillet. Add water, lemon juice, and 1 tsp. salt. Bring to boiling; reduce heat. Cover and simmer about 10 minutes or until fish flakes easily when tested with a fork. Cook green beans according to package directions; drain. In blender container combine milk, cheese, flour, butter or margarine, and ¼ tsp. salt. Cover and blend till ingredients are well-combined. Pour the mixture into medium saucepan. Cook and stir till thickened and bubbly; cook and stir 2 minutes more. Stir in sherry and beans. Drain fish; transfer to a warm platter. Spoon bean sauce over fish; sprinkle with almonds. If desired, garnish with parsley and lemon wedges.

FRIED MARINATED DOLPHIN

preparation time: 15 minutes
marinating time: 1 hour
cooking time: 10 minutes
serves: 12

Paula Taylor
Viking Maiden

12 Dolphin steaks
Juice of 5-6 lemons
Flour

Salt and pepper
Garlic powder

Marinate dolphin steaks in lemon juice at least 1 hour before cooking, this helps to cook the steaks, so actual frying time is reduced. Coat dolphin steaks with flour and seasonings. Fry in large pan in a butter-oil mix. Oil added to butter will keep the butter from burning. Serve garnished with lemon slices and Bernaise Sauce. *Serve with Fresh Mushroom Salad, White Rice with Almonds, fresh broccoli with butter and Apple/Blackberry Crumble.*

GOTES FISH IN HAWAIIAN STYLE

preparation time: 15 minutes
cooking time: 20 minutes
serves: 6

Pernilla Stahle
Adventure III

2 pkgs. flounder fillets, frozen
2 cups mayonnaise
2 Tblsp. Dijon mustard
1 Tblsp. sugar
1 can (16 oz.) pineapple bits and
    syrup

4 medium bananas
Almond flakes or slivers
1 Tblsp. curry
salt and pepper

Mix mayonnaise, mustard, sugar, curry, salt, pepper and pineapple syrup from can. Put defrosted fish (lightly fried if time) in oven dish. Spread pineapple bits and bananas over it. Pour over the mixture. Top with almond flakes. Bake until golden in color at 440° for 20 minutes. Note: Make it HOT Curry Pepper to cut the sweet taste! *Serve with curried rice, snow peas, green salad and Vanilla Parfait with Kiwi Fruit and Blackberries.*

GLAZED GRILLED SALMON

preparation time: 10 minutes
cooking time: 10 minutes
serves: 2

Jan Robinson
Vanity

2 (7 to 8 oz.) salmon steaks
    (about 3/4 inch thick)
3 Tblsp. (packed) dark brown sugar
4 tsp. prepared Chinese-style or
    Dijon mustard

1 Tblsp. soy sauce
2 tsp. rice vinegar

Prepare barbecue (Medium-high heat). Combine brown sugar, mustard and soy sauce in medium bowl; whisk to blend. Transfer 1 tablespoon glaze to small bowl; mix in rice vinegar and set aside. Brush 1 side of salmon steaks generously with half of glaze in medium bowl. Place salmon steaks, glazed side down, into barbecue. Grill until glaze is slightly charred, about 4 minutes. Brush top side of salmon steaks with remaining glaze in medium bowl. Turn salmon over and grill until second side is slighty charred and salmon is just opaque in center, about 5 minutes longer. Transfer salmon to plates. Drizzle reserved glaze in small bowl over salmon and serve. *Serve with steamed baby red potatoes and green beans.*

TERIYAKI-MARINATED SWORDFISH

preparation time: 15 minutes
marinating time: 2 hours
cooking time: 10-15 minutes
serves: 6-8

Pamela McMichael
Pieces of Eight

1-1/3 cup light soy sauce
2/3 cup medium-dry sherry
2 Tblsp. sugar
1 Tblsp. grated ginger root

2 cloves garlic, crushed to a paste
1 swordfish steak (3 to 4 lbs.), cut
    1-1/4" to 1-1/2" thick

Heat soy sauce, sherry, sugar, gingerroot and garlic over medium heat to boiling. Strain. Place sowordfish in strained soy marinade. cover and refrigerate 2 hours, turning occasionally. Grill 4" above coals, basting frequently, turning fish once until center is just opaque and outside is slightly charred, 5 to 8 minutes each side. Serve immediately. Don't forget to oil grill rack.

DESSERTS

BAKLAVA

preparation time: 45 minutes
cooking time: 1 hr.

Jane Mariott
Jan Pamela II

1½ lb. chopped walnuts
½ cup sugar
2 Tsp. ground cinnamon
1 Tsp. ground cloves
½ Tsp. freshly-grated nutmeg
1 lb. fillo pastry leaves
1 lb. melted butter

For the Syrup:
2 cups sugar
1½ cups water
3 small cinnamon sticks
Grated rind of 1 lemon
1 cup honey
¼ cup fresh lemon juice
3 Tblsp. brandy

Mix together the chopped walnuts, sugar and spices. Brush a large (16" x 11") pan with melted butter. Layer 6 fillo leaves in pan, brushing each with melted butter. Sprinkle ¼ of the walnut mixture over fillo. Top with 4 fillo leaves, brushing each with melted butter. Sprinkle with ¼ walnut mixture. Repeat layers twice more, using 4 fillo leaves for each layer. After last layer of walnut mixture, top with 6 fillo leaves, brushing each with melted butter. Cut baklava into diamonds, using a very sharp knife, dipping it in water to aid cutting. Bake at 350° until golden, about an hour. Pour cooled syrup over baklava as soon as they come out of the oven. Store without refrigerating at least eight hours before serving. After serving — go on a diet!

Syrup — Combine sugar, water, cinnamon sticks and lemon rind, bring to a boil and cook for about ten minutes. Strain, cool slightly, then add the honey, lemon juice and brandy.

APPLE KUCHEN

preparation time: 30 minutes
cooking time: 35 minutes
serves: 6

Lori Moreau
Alberta Rose

1 pkg. yellow cake mix
4 oz. butter
2 apples, sliced
½ cup sugar

Cinnamon
2 egg yolks
1 cup sour cream

Preheat oven to 350°. Mix together butter and cake mix. Mix until crumble. Pat lightly into bottom of a 9x13" pan, building up edges slightly. Bake 10 minutes. Combine sour cream and egg yolks, mixing well. Layer apple slices on warm crust. Sprinkle with sugar and cinnamon. Drizzle sour cream mixture over apple slices (will not cover completely). Bake for 25 minutes. This recipe makes a lot of kuchen, but leftovers make a super coffee cake for next morning's breakfast.

APPLESAUCE CARROT CAKE

preparation time: 10-15 minutes　　　　　　　　*Ann Glenn*
cooking time: 35-45 minutes　　　　　　　　　　*Encore*
serves: 8-12

1 pkg. Carrot 'n' Spice cake mix　　1 cup pecans or walnuts
1 jar (15 oz.) applesauce　　　　　　½ cup raisins
3 eggs

Preheat oven to 350°. Grease and flour well small individual Bundt pans
or regular Bundt pan. Blend cake mix, applesauce and eggs until moistened. Beat 2 minutes at highest speed. By hand, stir in nuts and raisins.
Pour into prepared pan. Bake at 350° for 35-45 minutes, or until toothpick inserted in the center comes out clean. Start toothpick test after 10
minutes for 12 x ½ cup Bundtletts, and after 15 minutes for 6 x 1 cup
Bundtlett pans. *Serve with Rum Cream.*

RUM CREAM

preparation time: 5 minutes
chilling time: 30 minutes
serves: 8-10

1 small package Jello Instant　　　¼ cup light rum
　　French Vanilla Pudding　　　　4 oz. Cool Whip
1 cup milk

Beat milk and pudding mix 2 minutes. Add rum and beat 1 minute. Fold
in Cool Whip. Refrigerate. Serve from a pretty pitcher or crystal bowl
with a ladle. Astonishingly simple, and wonderfully versatile. I have made
the same cream with Amaretto or Kahlua instead of the rum, and I suspect that Frangelico or Cointreau would be equally delightful.

BETTER THAN SEX CAKE

preparation time: 15 minutes　　　　　　　　*Terry Booey*
cooking time: 1 hour　　　　　　　　　　　　*Glen Mac*

1 box yellow cake mix　　　　　　　1 cup sour cream
　　(without pudding)　　　　　　　1 bar grated german chocolate
1 box instant vanilla pudding　　　1 (6 oz.) pkg. chocolate chips
½ cup oil　　　　　　　　　　　　　1 (6 oz.) pkg. butterscotch chips
½ cup water　　　　　　　　　　　　½ cup chopped pecans
4 eggs

Mix in order given, add eggs one at a time. Fold in all chocolate and nuts.
Put in bundt pan (greased). Bake 350° for 1 hour.

CARROT CAKE

preparation time: 30 minutes   *Launa Cable*
cooking time: 45 minutes   *Antiquity*

2½ cups flour
1½ tsp. soda
1½ tsp. cinnamon
2 tsp. baking powder
1 tsp. salt
2 cups sugar

1¼ cups crisco oil
4 eggs
2 cups grated carrots
2 cups crushed and drained
   pineapple
½ cup nuts

Sift dry ingredients, add sugar, oil and eggs. Mix well. Add carrots, pine-apple and nuts. Grease and flour pan. Bake at 350 for 45 minutes.
Cream Cheese Icing:

¼ lb. butter
8 oz. cream cheese

1 lb. confectioners sugar
1 tsp. vanilla

Mix all ingredients together, ice cake and eat . . . mmmm!

CARROT CONNECTION CAKE

preparation time: 15 minutes   *Sandy Culbert*
cooking time: 35-45 minutes   *Mystique*

1 box Carrot Cake Mix (Duncan
   Hines, Betty Crocker)
1 (16 oz.) can puree of chestnuts

2 oz. Rum, or any other liquor —
   optional

Prepare cake mix as per instructions on box, but adding the chestnuts. Also add about 2 oz. of rum or any other liquor you like. Cook a little slower and longer than directions read. Allow cake to cool slowly in oven. Frost very sparingly, as the cake is super rich. Whipped cream or sour cream dollops would be enough. This cake was a mistake, but it turned out so well I couldn't believe it. Actually, the chestnuts could do wonders for all sorts of cake mixes, as they do not have a heavy taste, but serve to give a lovely moist texture.

• *When cream becomes difficult to whip, add a few drops of lemon juice. Cream will whip quickly.*

AMARETTO CHEESECAKE

preparation time: 45 minutes　　　　　　　　　　*Chris Kling*
cooking time: 1 hour　　　　　　　　　　　　　　*Sunrise*
serves: 12

Graham Cracker Crust:

1½ cups graham crumbs
½ cup powdered sugar
½ stick butter, softened
**½ cup ground almonds (grind in
　food processor or blender)**

Filling:

2 (8 oz.) pkgs. cream cheese

1 cup sugar
4 eggs, separated
1 cup sour cream
1 tsp. vanilla
2 Tblsp. cornstarch
½ cup Amaretto

Preheat oven to 350°. Grease an 8-inch springform pan. Combine crust ingredients and press into pan to form crust. Beat the cream cheese until creamy. Gradually beat in the sugar, then beat in the egg yolks, sour cream, vanilla and cornstarch. In a separate bowl beat the egg whites until stiff and fold into cheese mixture. Pour into the prepared crust and drizzle the Amaretto throughout the cheese. Swirl around slightly with a spoon. Bake for 1 hour or until golden brown on top.

• *Freshen shredded coconut by soaking in fresh milk and a dash of sugar for a few minutes or place in a sieve and set over boiling water and steam until moist.*

CHOCOLATE CHEESE CAKE

preparation time: 15 minutes
cooking time: 60 minutes
chilling time: overnight
serves: 8

Maureen Stone
Cantamar IV

¾ cup sour cream
2 (8 oz.) pkgs. cream cheese,
   softened
1 cup sugar
6 oz. semi-sweet chocolate pieces
   (melted)

4 eggs
1 tsp. almond flavoring
1 9-inch graham crust

Melt chocolate pieces with almond flavoring, keep warm. In large bowl, beat softened cream cheese. Blend in sour cream and eggs one a time. Add melted chocolate until well mixed. Pour into cracker crust and bake 1 hour at 325°. Chill several hours or overnight.

EGGNOG CHEESECAKE

preparation time: 15 minutes
cooking time: 50 minutes
serves: 6

Nancy Thorne
Gullviva

Crust:

1¾ cups graham cracker crumbs
2½ Tblsp. melted butter
2 Tblsp. sugar

Butter a 9 inch springform pan. Combine all ingredients and blend well. Pat mixture into pan and chill. Preheat oven to 350°.

Filling:

20 oz. cream cheese, softened
⅓ cup half and half
¼ cup whipping cream
¾ cup sugar
1½ tsp. vanilla
3 eggs

2 egg yolks
2½ Tblsp. dark rum
2 Tblsp. Cognac
Fresh grated nutmeg
Whipped Cream (optional)

Beat cream cheese until smooth. Gradually add half & half and whipping cream. Add sugar and vanilla. Mix well. Add eggs and extra yolks one at a time. Stir in rum and Cognac. Pour batter into crust. Grate nutmeg on top. Bake 45 to 50 minutes. Garnish with whipped cream if desired. *DO NOT EAT MORE THAN 3 PIECES AT A TIME, AS IT IS QUITE FAT-TENING.*

CHEESECAKE SUPREME

preparation time: 30 minutes   *Debbie Olsen*
cooking time: 1 hour 10 minutes   *Kingsport*
serves: 8

1¼ cup graham cracker crumbs   1¾ cup sugar
¼ cup sugar   3 Tblsp. flour
¼ cup melted butter   ¼ cup heavy cream
5 (8 oz.) pkg. cream cheese, room   5 whole eggs plus 2 egg yolks
  temperature

Preheat oven to 375° to bake crust. Work graham cracker crumbs, sugar and melted butter together. Press mixture over bottom and sides of a buttered 9-inch springform pan. Bake at 375° for 8 minutes. Beat cream cheese until smooth. Add sugar, flour, cream, eggs and egg yolks. Beat just until mixture is smooth. Pour into crust and bake at 500° for 10 minutes. Reduce oven temperature to 200° and continue baking for one hour.

4 LAYER DELIGHT

preparation time: 30 minutes   *Terry Booey*
cooking time: 15-20 minutes   *Glen Mac*

1 cup flour   2 pkgs. instant pudding (1 vanilla
1 stick butter, softened   and 1 chocolate)
½ cup pecans, chopped   3 cups cold milk
1 cup cool whip   1-2 cups cool whip
1 (8 oz.) pkg. cream cheese   Garnish: shaved chocolate or
1 cup powdered sugar   graham cracker crumbs

Layer I: Combine flour, butter and pecans; press in a 13x9 pan. Bake
           15-20 minutes at 350°. Cool.

Layer II: Blend well 1 cup cool whip, cream cheese and powdered sugar.
           Spread carefully over Layer I.

Layer III: Mix pudding and milk. Beat two minutes and spread over layer
           II.

Layer IV: Spread 1-2 cups of cool whip on top. Garnish. This sounds involved, but it's easy and guests love it!
           Liz Thomas on Raby Vaucluse uses ½ cup oatmeal in place of ½ cup pecans. Layer III she uses 2 pkgs. instant chocolate pudding. *For a change try a different pudding flavor; butter pecan, lemon, French vanilla.*

CHERRY CHOCOLATE CAKE

preparation time: 15 minutes
cooking time: 45 minutes
serves: 8

Lori Moreau
Alberta Rose

1 box chocolate cake mix
1 12 oz. can cherry pie filling
2 eggs (lightly beaten)

Combine ingredients. Bake in greased 10 x 13″ pan at 350° for 45 minutes.

Chocolate Frosting:

1 cup sugar
1 6 pz. pkg. chocolate chips

5 Tblsp. butter
⅓ cup milk

Bring combined sugar, butter and milk to a boil. Cook 3 minutes. Add chocolate chips. Cool and spread on cooled cake.

CHOCOLATE CHOCOLATE CAKE

preparation time: 5 minutes
cooking time: 1 hour
serves: 10

Abbie T. Boody
Ductmate I

1 pkg. chocolate cake mix
1 pkg. chocolate jello instant
pudding
1 cup chocolate morsels

½ cup oil
4 eggs
½ cup water
2 oz. powdered sguar

Mix all ingredients. Pour into greased bundt pan. Bake for 1 hour.

DUMP CAKE

preparation time: 15 minutes
cooking time: 1 hour 20 minutes
serves: 8-10

Betsi Dwyer
Malia

1 (16 oz.) can crushed pineapple
1 (16 oz.) can cherry pie filling
1 box yellow or white cake mix

1 cup pecans, chopped
½ lb. butter

Dump first 3 ingredients into well-buttered angel food cake pan. (A 9x13 pan will also work.) Sprinkle pecans on top and dot with pats of butter. Bake at 350°. *Serve hot or cold with whipped cream or ice cream.*

CHOCOLATE FLOP

preparation time: 25 minutes   *Lori Moreau*
cooking time: 35-40 minutes   *Alberta Rose*

¾ cup dark corn syrup
⅓ cup firmly packed dark brown
   sugar
¼ cup butter
⅛ tsp. salt
⅔ cup chopped pecans or walnuts

1 pkg. fudge cake mix (1 layer
   size)
1 egg
¾ cup water
¾ cup chocolate chips
Whipped cream

Preheat oven to 350°. Combine corn syrup, brown sugar, butter and salt in a saucepan and heat until butter melts. Remove from heat, stir in pecans. Pour into a greased 9x9x2″ pan and cool. Prepare cake mix with egg and water. Sprinkle chocolate chips over pecan mixture. Pour in batter. Bake 35-40 minutes. Turn out onto serving platter immediately. Serve warm with whipped cream.

FRENCH CHOCOLATE CAKE

preparation time: 15 minutes   *Pernilla Stahle*
cooking time: 30-35 minutes   *Adventure III*
serves: 8

1½ sticks butter or margarine
7 oz. semisweet chocolate
¾ cup sugar
3 egg yolks

4 oz. finely chopped hazelnuts
½ cup flour
½ tsp. Nescafe Instant coffee
3 egg white

Melt butter in saucepan. Add chocolate pieces. Add egg yolks and sugar, mix well. Separately mix nuts, flour, coffee and add to the saucepan. Whip egg whites until you can hold bowl upside down without it falling out, carefully blend it in. Bake immediately in 8 inch, greased pan at 400°, 30-35 minutes. Note: it should be creamy in the middle and heavy. *Serve with whipped cream.*

• *To cut a meringue-topped pie, grease the knife and the meringue never tears.*

HARVEY WALLBANGER CAKE

preparation time: 20 minutes   *Launa Cable*
cooking time: 55 minutes   *Antiquity*
cooling time: 1 hour

1 box yellow cake mix
1 box of vanilla pudding
1 scant cup oil
4 eggs
¼ cup vodka
¼ cup Galliano
⅔ cup orange juice

Topping:

1 stick butter
1 cup sugar
¼ cup orange juice
¼ cup Galliano

Mix cake and pudding mix, add other ingredients. Beat well about 2 minutes. Pour into buttered Bundt pan. Bake at 340° for 55 minutes. Cake is done when it springs back from fingers. Cool 30 minutes in pan then pour on half the topping. Let cool another ½ hour then loosen. Turn out on cake plate. put topping on bottom side.

Topping:

Mix together all ingredients and boil for 4 minutes.

MOSAIC ICE-CREAM CAKE

preparation time: 15 minutes   *Carol Muller*
freezing time: 3-4 hours   *S'Oublier*
serves 6-8

1 qt. vanilla ice cream
1 (6 oz.) can concentrated orange
   juice

1 pound cake (frozen or fresh)

Select stainless (3 qt.) bowl. Break pound cake into bite-size pieces and cover bottom of bowl. Drizzle orange juice concentrate over pieces of cake. Cover with layer of ice cream. Add another layer of pound cake, drizzle more orange juice, another layer of ice-cream until all ingredients used. Freeze 3-4 hours. To free from bowl, run knife around sides and place in bowl of hot water momentarily. Turn out on platter. If desired, garnish with whipped cream and fruit in season. Cut slices revealing mosaic pattern. *If boat freezer won't freeze solid, serve mixture in bowls. for delicious "Paradise Pudding."*

• *To remove the white membrane from oranges easily, soak them in boiling water for a few minutes before you peel them.*

RAVE REVIEW

preparation time: 15-20 minutes
cooking time: 35 minutes

Terry Booey
Glen Mac

Cake
1 pkg. yellow cake mix
1 pkg. instant coconut pudding
1⅓ cups water
4 eggs
¼ cup oil
2 cups coconut
1 cup nuts (toasted)

Coconut Icing:
4 Tblsp. butter
2 cups coconut
1 8 oz. pkg. cream cheese
2 tsp. milk
1 box powdered sugar
1 tsp. vanilla

Cake — Combine and mix well. Put into 3 greased and floured pans. Bake 35 minutes at 350°.
Icing — Cream butter and cream cheese. Add to sugar, coconut, milk and vanilla. Mix well and ice between layers.

SOUR CREAM COFFEE CAKE

preparation time: 25 minutes
cooking time: 1 hour

Launa Cable
Antiquity

4 cups flour
1 cup butter
2 cups sugar
2 tsp. baking powder
2 tsp. vanilla
1 tsp. salt
2 tsp. soda
4 eggs
2 cups sour cream

Topping

1 cup sour cream
1 tsp. vanilla
½ cup chopped nuts
1 tsp. cinnamon
¼ cup sugar
4 Tblsp. brown sugar

Cream butter and sugar, add eggs and beat. Sift dry ingredients together and add alternately with sour cream. Stir in vanilla. Pour half of batter in greased bundt pan. Sprinkle on half the topping, add remaining batter and sprinkle with rest of topping. Bake at 350° for 1 hour.

• *Melt chocolate chips in double boiler over hot, not boiling, water. When boiled, chocolate will not harden properly and will have a white cast.*

SIMPLY SINFUL

preparation time: 15 minutes   Harriet Beberman
serves: 4-6   Avatar

1 Pound cake   **2 bananas, sliced**
1 jar orange marmalade   **Whipped cream**
½ cup rum or amaretto

Slice pound cake lengthwise into 3 or 4 layers. Combine marmalade and rum. Spread on first layer of cake. Add layer of bananas followed by a layer of whipped cream. Place a layer of cake on top and repeat process until all layers are filled. Frost cake with remaining whipped cream and refrigerate until ready to serve. Alternatives: strawberry preserves, peach or pineapple instead of orange. Sprinkle almonds, pecans or walnuts on top.

WINE CAKE

preparation time: 25 minutes   Lori Moreau
cooking time: 45-55 minutes   Alberta Rose
serves: 8-10

1 (18 oz.) yellow cake mix   **¾ cup vegetable oil**
1 (4½ oz.) instant vanilla pudding **2½ cups confectioners sugar**
mix   **4 eggs**
1¼ cup "inexpensive" sherry,
divided

Preheat oven to 325°. Combine cake mix, pudding mix, ¾ cup sherry and oil. Mix well. Add eggs, one at a time. Beat VERY well. Pour batter into greased and floured 10″ tube pan. Bake 45-55 minutes. Cool 10 minutes in the pan. Make a glaze with ½ cup sherry and confectioner's sugar. Poke cake bottom with a straw. Pour ½ the glaze over. Invert cake onto serving plate and repeat. This cake is best when allowed to sit all day uncovered.

• *Small marshmallows make unique candle holders for a birthday cake and the melted wax will not run down into the frosting.*

CREME DE MENTHE PUDDING IN A CLOUD

preparation time: 15 minutes
chilling time: 1-2 hours
serves: 4-6

Shirley Fortune
Maranatha

1 pkg. INSTANT chocolate
   pudding
1¾ cup milk, cold
Large tub Cool Whip

¼ cup creme de menthe (green)
A chocolate bar to shave curls to
   garnish

To prepare pudding: Pour 1¾ cup milk into a bowl, add pudding mix
and creme de menthe. Whip until thickened — it can be done by hand!
Chill until serving time. *To serve: Line bottoms of champagne glasses with
a generous amount of Cool Whip and fill to the top with pudding. Add a
dollop of Cool Whip to the center and garnish with chocolate curls.*

FRANGELICO VELVET

preparation time: 25 minutes
chilling time: 1½-2 hours
serves: 6-8

Shirley Fortune
Maranatha

4 eggs, separated
1 Tblsp. sugar
6 oz. pkg. chocolate chips

¼ cup boiling water
¼ cup Frangelico Liqueur
Chopped hazelnuts or peanuts

Beat egg whites until foamy, gradually adding sugar until stiff, set aside.
Whirl chocolate chips in blender until finely grated. Add boiling water and
blend until chocolate melts. Add egg yolks and liqueur and blend until
smooth. Gently fold chocolate mixture into whites. Spoon into dessert
dishes or champagne glasses and chill until set. Garnish with chopped
nuts around edges and one whole nut at the center. *Note: serving in
glasses makes a more stunning presentation!*

CHERRY STARS

preparation time: 10 minutes
cooking time: 10-15 minutes
serves: 6

Patty Dailey
Western Star

Vanilla wafers
1 lb. can cherry pie filling

1 (8 oz.) cream cheese

Spread ¼-inch of cream cheese on a vanilla wafer. Top with 1 tsp. cherry
pie filling. Bake at 350° for 10-15 minutes.

POTS OF CHOCOLATE

preparation time: 30 minutes   *Jean Thayer*
chilling time: 2 hours   *Paradise*
serves: 6

4 oz. semi-sweet chocolate   1 Tblsp. Grand Marnier
2 Tblsp. butter   4 egg yolks, beaten
Juice and grated rind of 1 orange 3 egg whites

Melt chocolate, butter and orange juice in the top of a double boiler. Remove from heat and add grated rind and Grand Marnier. Allow chocolate to cool slightly and add egg yolks stirring well. Beat egg whites till stiff and gently fold into chocolate. Spoon into 6 dessert cups and chill for at least 2 hours.

CHOCOLATE-ORANGE BROWNIES

preparation time: 30 minutes   *Pamela McMichael*
cooking time: 18 minutes   *Pieces of Eight*

Unsalted butter for baking pan   ½ tsp. vanilla
4 oz. bittersweet chocolate,   Pinch of salt
    chopped   ½ cup all-purpose flour
4 Tblsp. unsalted butter   ¼ tsp. baking powder
6 Tblsp. sugar   2 tsp. Cointreau
1 egg beaten, slightly   ½ cup heavy cream
1 tsp. orange zest

Heat oven to 350°. Line bottom of 8″ square baking pan with foil. Butter foil and sides of baking pan. Melt chocolate and 4 Tblsp. butter in small heavy saucepan. Cool 2 minutes. Stir in sugar till blended. Stir in egg, orange zest and vanilla till blended. Sift together flour, baking powder and salt. Blend into chocolate mixture. Scrape into prepared pan. Bake in oven till edges of brownies pull away from sides of pan, about 18 minutes. Cool 5 minutes. Invert pan and peel off foil. Divide brownies into 4 equal squares. Sprinkle with ½ tsp. Cointreau. *At serving time beat heavy cream until soft peaks form. Top each brownie with a spoonful of cream.*

• *Cool melted chocolate before adding to mixture containing egg whites or whites will flake.*

LIGHT CHOCOLATE PUDDING

preparation time: 25 minutes (plus cooling)   *Pamela McMichael*
cooking time: 15 minutes   *Pieces of Eight*
serves: 4

2 cups milk
½ cup sugar
2 Tblsp. cornstarch
Pinch salt
3 oz. bittersweet chocolate broken
    into ½" pieces

½ cup heavy cream
1 Tblsp. orange and nut flavored
    liqueur
1 tsp. vanilla

Scald milk and vanilla. Remove from heat. Stir sugar, cornstarch and salt in second saucepan till blended. Gradually stir in hot milk and vanilla; stir in chocolate. Cook, stirring till chocolate has melted and mixture has thickened; about 1 minute. Pour into custard cups and cool. *At serving time, whip cream and fold in liqueur. Serve pudding with whipped cream on side.*

COLD LOVE

preparation time: 25 minutes   *Jane Marriott*
marinating time: 2-3 hours   *Jan Pamela II*
cooking time: 15 minutes
chilling time: 2-3 hours
serves: 6

A selection of fresh fruit as
    colourful as possible peeled
    and chopped
⅔ cup dark rum
1-2 Tblsp. sugar
Custard:
4 egg yolks, beaten

½ tsp. freshly-grated nutmeg
2½ cups milk
½ cup sugar
½ tsp. vanilla extract
⅔ cup heavy cream, whipped
1 envelope unflavored gelatin

Sprinkle the fruit with sugar and soak it in the rum for 2-3 hours. Make the custard in a heavy bottomed pan, or in a double boiler and over hot, not boiling, water. Combine the milk, beaten egg yolks, sugar and nutmeg. Stir continuously until mixture becomes thick. Add vanilla. Soften the gelatin in a Tblsp. of warm water and stir into the custard until completely dissolved. Mix together the fruit (drained) and custard and fold in the whipped cream. Chill for several hours before serving. *This dessert looks very pretty served in tall glasses with a strawberry or cherry for a garnish on top of each serving.*

CHOCOLATE DESSERT CUPS

preparation time: 20 minutes   Gill Case
chilling time: 1½ hours   M/V Polaris
serves: 4

6 oz. chocolate chips   1 Tblsp. rum
2 Tblsp. butter   1 Tblsp. coffee essence or Kahlua
3 eggs   Grated chocolate to decorate

Put chocolate and butter in a bowl placed over a pan of hot water. Leave
to dissolve, stirring constantly. Separate eggs. Beat yolks into melted
chocolate with rum and coffee essence. Whisk whites stiffly, fold into
chocolate mixture. Pour into small glasses and refrigerate for at least 1½
hours till set. Decorate with grated chocolate just before serving.

CONFETTI COOKIES

preparation time: 20 minutes   Launa Cable
cooking time: 15 minutes   Antiquity
makes: 2-3 dozen

2⅓ cups all-purpose flour   1 cup shortening
1 cup granulated sugar   2 eggs
½ cup firmly packed brown sugar   1 tsp. vanilla
1½ tsp. baking soda   1 cup cut-up gum drops
1 tsp. salt

Add all ingredients, except gum drops and mix well. Stir in gum drops.
By teaspoonfuls drop onto greased baking sheet. Cook at 350° for 15
minutes.

HELLO DOLLY'S

preparation time: 10 minutes   Phoebe Cole
cooking time: 34-45 minutes   Barefoot
serves: 6-8

1 stick butter, melted   1 cup semi-sweet chocolate bits
1 cup graham crackers crumbs   1 cup chopped walnuts
1 cup coconut flakes   1 can Eagle Brand condensed
                                           milk

Layer ingredients in brownie pan, in order given. Bake at 350° until firm,
35-45 minutes.

LUMBER-JACK COOKIES

preparation time: 25 minutes   *Launa Cable*
cooking time: 12-15 minutes   *Antiquity*
makes: 4 dozen

1 cup sugar
1 cup shortening
1 cup dark molasses
2 eggs
4 cups sifted flour

1 tsp. salt
2 tsp. cinnamon
1 tsp. ginger
1 tsp. soda
Sugar for rolling

Cream together sugar and shortening. Add molasses and eggs. Mix well. Sift together dry ingredients and add. Put ¼ cup sugar into small bowl. Dip fingers into sugar — pinch dough (tablespoon size) roll in sugar. Place balls on greased cookie sheet. Bake at 350° for 12-15 minutes.

RUM BALLS

preparation time: 30-40 minutes   *Betsi Dwyer*
ripening time: 24 hours   *Malia*

2 Tblsp. dark corn syrup
2½ cups vanilla wafers, crushed
    (almost a 12 oz. box)
1¼ cups chopped pecans or
    walnuts

2 Tblsp. cocoa, unsweetened
½ cup rum
4 Tblsp. cocoa
½ cup confectioners sugar

Sift 4 Tblsp. cocoa and confectioners sugar together. Combine crushed wafers, nuts, cocoa and corn syrup. Add enough rum to make a firm dough. Roll dough into balls in the palm of your hands. Roll balls in sifted cocoa and sugar. Store tightly covered for at least 24 hours to ripen.

BANANA CHOCOLATE CREAM

preparation time: 5 minutes   *Paula Taylor*
serves: 12   *Viking Maiden*

2 pkgs. (6⅛ oz.) instant chocolate
    pudding
4 bananas

Whipped cream
Chocolate bits

Mix instant pudding as per instructions on packet. Slice bananas into mix. *Serve in individual bowls topped with whipped cream and chocolate bits.*

ALMOND CREAM

preparation time: 15 minutes
chilling time: 4 hours
serves: 6

Michelle Sutic
Harp

1 pkg. gelatin
1 cup water
¼ cup sugar
2 cups half 'n half

2 tsp. almond extract
1 can mandarin oranges
Whipped cream for garnish

Sprinkle gelatin on water and dissolve over low heat. Add sugar and stir until dissolved. Add almond extract. Cool. When cool add half and half and mix well. Pour into glass serving bowl and refrigerate approximately 4 hours until set. *Just before serving decorate top with drained mandarins and whipped cream.*

AMARETTO CREAM

preparation time: 5 minutes
chilling time: 1 hour
serves: 6-10

Ann Glenn
Encore

1 (3 oz.) jello Instant French
    Vanilla pudding
1 cup milk

¼ cup Amaretto (or light rum or
    kahlua, or other favorite
    liqueur)
4 oz. Cool Whip

Combine pudding mix and milk and beat 2 minutes. Add liqueur and beat 1 minute. Fold in Cool Whip gently. Chill until time to serve. Particularly recommended are Amaretto Cream over frozen peaches, and Rum Cream with Applesauce Carrot Cake.

DESSERT CREAM

preparation time: 3 minutes
serves: 6-8

Marty Peet
Grumpy III

4 oz. container of Cool Whip
3 Tblsp. Kahlua, Cointreau or
    Drambuie

Nutmeg
Cinnamon

Mix well — no one will know it is Cool Whip. *Serve over desserts, in coffee or over breakfast breads or pancakes.*

GUAVA CREMA

preparation time: 20 minutes　　　　　　　*Silvia Kahn*
chilling time: 2 hours　　　　　　　　　　*Antipodes*
serves: 6

2 cans guava in heavy syrup　　　1 tsp. nutmeg
　(guayaba)　　　　　　　　　　1 Tblsp. lime juice
1 cup cream cheese　　　　　　　3 tsp. brown sugar
1 shot rum　　　　　　　　　　　2 egg yolks

Wash heavy syrup from guavas. Chop in pieces. In mixing bowl combine all other ingredients. Mix till creamy. Stir in guavas and chill for 2 hours. *Serve in dessert cups individually. Decorate with piece of lime.*

ORANGE SAUCE CREPES

preparation time: 15 minutes　　　　　　*Terry Boudreau*
cooking time: 20 minutes　　　　　　　　*High Barbaree*
serves: 6

⅓ cup orange juice　　　　　　　½ tsp. grated orange peel
¼ cup butter or margarine　　　　¼ cup orange flavored liqueur
2 Tblsp. sugar　　　　　　　　　12 crepes

In 10-inch skillet or chafing dish over low heat, heat juice, butter, sugar and orange peel until butter melts. Fold crepes in quarters, arrange in sauce; heat through. In a very small saucepan over medium heat, heat liqueuer until hot; remove from heat. Ignite liqueur with match, pour flaming liqueur over crepes. *When flame dies down, serve on warm dessert plates.*

EASY ECSTASY

preparation time: 10 minutes　　　　　　　*Jan Robinson*
chilling time: 2 hours　　　　　　　　　　*Vanity*
serves: 6-8

1 can sweetened condensed milk　　1 can strawberry pie filling
1 large can crushed or chunk　　　1 (9 oz.) container whipped
　pineapple in juice　　　　　　　　topping

Mix all ingredients thoroughly. Refrigerate until ready to serve. *Serve in decorative glasses and top with a fresh strawberry.*

STRAWBERRY CREPES

preparation time: 20 minutes   *Launa Cable*
cooking time: varies   *Antiquity*
serves: 4-6

Crepes:

1 cup Bisquick
2 eggs
¾ cup milk
1 tsp. vanilla

Filling:

6 oz. softened cream cheese
2 Tblsp. sour cream
2 Tblsp. sugar
1 tsp. vanilla
1 tub of frozen strawberries with
   syrup

Set aside a pan of boiling water with tin foil on top with slits cut through foil to allow steam through. Place a cover over that (place cooked crepes on top to keep warm). In small stick proof 5" fry pan, heated, pour 1 large serving spoonfull of crepe mixture into pan, slowly swirl pan for all mix to cook before flipping to cook other side, repeating until all of mixture is cooked up into crepes. Place 1½ Tblsp. of filling into middle of crepe adding a few strawberries with syrup. Fold over sides, top again with strawberries and syrup. Garnish with orange slices.

ENGLISH POUNDS

preparation time: 10 minutes   *Gay Thompson*
cooking time: 5-10 minutes   *Satori*
serves: 8

1 lb. pound cake
1 lb. can pears
4 oz. jar fudge topping

2 ozs. Grand Mariner
Maraschino cherries
Whipped cream

Slice a 1 inch thick piece of pound cake for each person and place on ungreased cookie sheet. Place one pear half on each cake slice. Bake at 300° for 5 minutes or until warmed through. Meanwhile, warm fudge topping and Grand Marinier in a saucepan. Place cake and pears on individual serving dishes. Top with sauce, then whipped cream and finally a cherry. *Serve immediately.*

• *Fresh pineapple should be cooked when used with a gelatin base salad.*

FRESH FRUIT FLAN

preparation time: 20 minutes (plus cooling)   Shirley Fortune
cooking time: 20 minutes   Maranatha
serves: 6-8

1 cup sifted cake flour
1 tsp. baking powder
1 pinch salt
¼ cup softened butter
½ cup sugar
2 large egg yolks, beaten
⅓ cup + 1 Tblsp. milk
½ tsp. vanilla
¼ tsp. lemon rind
Fluted flan cake pan
Fresh fruits like kiwi, bananas and
    strawberries

Filling:
1 cup milk
1 cup Cool Whip
1 small pkg. INSTANT vanilla
    pudding

Glaze:
Melted apple jelly

Grease and flour flan pan well. Cream butter until light and add sugar, beat in yolks until fluffy. Combine flour, baking powder, salt, and add alternately with milk until combined. Add vanilla and rind and pour into pan and bake for 20 minutes at 350°. Remove from pan a few minutes after taking from oven and cool. Prepare filling while baking cake by combining all ingredients. Fill the indented center of cake with filling when cake is cooled. Let set before arranging fruit on top. Once set, arrange cut fruit in an attractive pattern such as bananas diagonally sliced over most of ⅔ top. In the ⅓ remaining section make a ring of overlapping kiwi slices. Then cut 1 heart shaped large strawberry lengthwise and arrange as a flower in center of kiwi slices. Glaze with warm jelly. Leftovers freeze well. A truly elegant and attractive dessert.

APPLE AND BLACKBERRY CRUMBLE

preparation time: 10 minutes   Paula Taylor
cooking time: 1-1½ hours   Viking Maiden
serves: 12

2 cans (15 oz.) blackberries
4-5 apples
¼-½ cup sugar

4 oz. butter
2 oz. sugar
8 oz. flour

Combine blackberries, sliced apples and sugar to taste. Heat oven to 400°. Place apples and blackberries in greased deep dish. Mix flour, sugar and butter into a crumble, place over fruit and cook for 1-1½ hours until crumble is brown. Cool and serve topped with whipped cream.

QUESILLO FLAN CARAMELO

preparation time: 40 minutes   Silvia Kahn
cooking time: 1 hour   Antipodes
cooling time: 5 hours

Custard:
3 eggs, separated
1 (14 oz.) can sweetened
**   condensed milk**
1½ cups whole milk
2 tsp. vanilla

2 tsp. grated coconut

Caramel:

1 cup white sugar
½ cup water

Warm canned and whole milk over low heat. Remové from flame and add egg yolks, coconut and vanilla. Set aside. Heat sugar and water in skillet stirring constantly until liquid turns caramel color, about 15 minutes. Turn carmel into pirex bowl and rotate, covering sides. Beat egg whites and fold into egg yolk and milk mixture. Pour into caramelized bowl. Sit bowl in pan of boiling water and put in oven at 350° for 1 hour. Cool at least 5 hours before serving.

BANANA BLAST

preparation time: 10 minutes (plus cooling)   Sandy Culbert
cooking time: 20-30 minutes   Mystique
serves: 6-8

2 cups flour
¾ cup sweet butter
Dash salt
3 Tblsp. confectioners sugar

3-4 bananas
½-¾ cup Cointreau
Whipped cream, arbitrary amount

Cut flour, sugar and salt into the butter as for any pie crust. Pat the dough into a shallow cake pan (8-inch type). Do not roll this dough. Cook at 300° for 10-15 minutes. Take out of oven and pat the dough up against the sides where it has sagged. Return and cook till it appears to be firm (about 15 minutes more). Cool pie shell. When ready, fill with the bananas soaked liberally in Cointreau and smothered in whipped cream. What's left over is great in the morning!

BANANAS BAREFOOT

preparation time: 5 minutes
cooking time: 15 minutes
serves: 4

Phoebe Cole
Barefoot

3 oz. Grand Marnier
½ cup sugar
¼ lb. butter

4 bananas
Ice cream — vanilla
Cinnamon

Put Grand Marnier, butter and sugar in large skillet. Stir at high heat until it starts to carmelize, cut bananas lengthwise, then peel and sauté until golden. Put on top of ice cream and spoon sauce over ice cream. Sprinkle with cinnamon. Note: very low in calories!!

BANANAS FOSTER

preparation time: 10 minutes
cooking time: 20 minutes
serves: 8

Betsi Dwyer
Malia

8 ripe bananas
½ cup 151 proof rum
¼ cup Grand Marnier

¼ cup B&B brandy
½ cup brown sugar
¼ lb. butter

Peel bananas and slice in half from poles. In a greased pan, place bananas and add ¼ cup rum, Grand Marnier, B&B, butter pats and ¼ cup brown sugar. Bake 20 minutes at 300°. Drain sauce and serve separately. Add remaining ¼ cup rum and ¼ cup brown sugar. *Light and serve flaming. An elegant, but simple dessert!*

MOUNT GAY BANANAS

preparation time: 5 minutes
cooking time: 5 minutes
serves: 6

Jean Thayer
Paradise

4 Tblsp. unsalted butter
⅓ cup firmly packed brown sugar
6 bananas, peeled, halved
    lengthwise and crosswise

½ tsp. cinnamon
⅓ cup Mt. Gay
2 Tblsp. lime juice

In large skillet over medium-high heat, melt butter. Stir in sugar and cinnamon and cook for 1 minute. Add bananas, stirring gently to coat. Pour in rum and ignite, shaking skillet till flames die. Stir in lime juice and serve.

RUBY BANANAS

preparation time: 10-15 minutes
cooking time: 15-20 minutes
serves: 6-8

Ann Glenn
Encore

1-1½ bananas per serving
1 can (16 oz.) whole berry
   cranberry sauce

Lemon juice
Cinnamon
Sweetened whipped cream

Choose a shallow, oven-to-table dish. Slice bananas lengthwise and cross-wise. Arrange in the dish in a single layer, if possible. Sprinkle with lemon juice. In a separate bowl, mash the cranberry sauce, then spread evenly over the bananas. Sprinkle with cinnamon. Heat 15-20 minutes in 250° oven. (Doesn't really require cooking, can be warmed through by placing in preheated oven.) *Serve individually, topped with whipped cream. The contrasts, between sweet and tart, hot and cold, are what make this dessert outstanding.*

COMMENT: This low calorie 'just fruit' dessert is usually my first night's offering. As the whipped cream makes its way around the table, I can judge who really meant it when they remarked the preference sheet "Prefer NOT to be tempted by breads and desserts every night"!!

OREGON BOYSENBERRY CUSTARD

preparation time: 15 minutes
cooking time: 20-30 minutes
serves: 4

Gay Thompson
Satori

2 eggs, slightly beaten
⅓ cup sugar
½ tsp. vanilla
1½ cups heated milk

1 (1 lb.) can Oregon
   boysenberries, blackberries or
   blueberries
¼ teaspoon nutmeg

Combine eggs, sugar and vanilla. Mix thoroughly. Add heated milk beating constantly. Mix in berries. Pour in a 2 quart casserole dish and sprinkle with nutmeg. Bake at 350° for 20-30 minutes (does not have to be baked in a dish of water in oven). It will be slightly soft in the center but will set as it cools.

WARM FRUIT DESSERT

preparation time: 25 minutes
cooking time: 10 minutes
serves: 6

Lori Moreau
Alberta Rose

(Fruit may be canned or fresh)
1 cup pineapple chunks
1 cup orange segments, halved
2 small bananas, sliced
¾ cup shredded coconut
¼ cup butter

¾ cup confectioners sugar
1 Tblsp. cornstarch
3 Tblsp. lemon juice
2 Tblsp. orange rind
⅓ cup orange juice

Melt butter in medium saucepan. Add lemon juice and orange juice. Gradually add cornstarch, blending well. Add sugar and orange rind. Cook till thick. Add fruit and heat through.

Fluffy Cheese Topping:

½ cup heavy cream
8 oz. cream cheese, softened
¼ cup confectioners sugar

⅛ tsp. nutmeg
2 tsp. grated orange rind

Beat cream until stiff. Beat cream cheese, sugar and spices. Add to whipped cream. *Serve on top of warm fruit.*

PEAR OR APPLE SUPREME

preparation time: 15 minutes
cooking time: 10 minutes
serves: 4

Kandy Popkes
Wind Song Oregon

2 fresh pears or apples
8 Tblsp. brown sugar
1 cup Jack Daniels
6 Tblsp. butter

Pound Cake
1 pint cream, whipped
Cinnamon

Peel, core and thinly slice pears, or apples. Melt butter, brown sugar and cinnamon in large shallow pan, preferably teflon. Saute fruit in butter mixture. Heat Jack Daniels in a separate saucepan; when hot pour over fruit. Reserve a tablespoon of JD to light over flame. Light the fruit mixture and stir until flame is out. In each serving bowl, place a piece of pound cake, top with flambe fruit mixture and whipped cream.

• *Sift powdered sugar over top of cake before frosting to prevent icing from running off the cake.*

BROILED GRAPEFRUIT AND GRAPES

preparation time: 25 minutes   *Pamela McMichael*
cooking time: 3-4 minutes   *Pieces of Eight*
serves: 4

2 medium-size grapefruit   ¼ cup light brown sugar
1 cup seedless green grapes,   ½ tsp. ground cinnamon
   halved
2 Tblsp. unsalted butter

Remove peel and pits from grapefruit, slice flesh into sections. Combine
grapefruit and grapes in 8x6x1½" broiler-proof dish. Cut butter into small
pieces and scatter over top. Mix brown sugar and cinnamon in bowl.
Sprinkle over fruit. Broil 3 to 4 minutes until butter-sugar topping is melted.

VERY SPECIAL GRAPES

preparation time: 15 minutes   *Cheryl Anne Fowler*
chilling time: 2 hours   *Vanity*
serves: 4

2 pounds grapes (seedless green, if   ½ cup Grand Marnier (or to taste)
   available)   ½ cup light brown sugar
1 cup sour cream

Combine sour cream and Grand Marnier. Add grapes and toss until coated. Sprinkle brown sugar on top. Chill for at least two hours. *Serve in
glasses.*

MELON BALLS

preparation time: 15 minutes   *Harriet Beberman*
chilling time: 2-4 hours   *Avatar*
serves: 4-6

1 cantelope melon   2 Tblsp. orange concentrate
1 honeydew melon   2 Tblsp. triple-sec or orange
¼ cup Karo white syrup   liqueur

Half melons, remove seeds and make melon balls with melon ball scoop.
Combine Karo syrup, orange concentrate, and liqueur. Pour over melon
balls and refrigerate for several hours, stirring occasionally. *Great as a
first course for breakfast, an appetizer, or a dessert.*

FRESH PAPAYA TROPICAL

preparation time: 5 minutes   *Silvia Kahn*
serves: 6   *Antipodes*

1 big papaya   **Nutmeg**
Lime juice   **Rum**
Brown sugar   **Condensed milk (sweet)**

Peel the papaya, discard seed and chop. Arrange papaya in dessert bowls. Sprinkle with sugar, nutmeg, rum and lime juice. Top with sweetened condensed milk, if desired.

PEAR-CRANBERRY CRISP

preparation time: 20 minutes   *Kathleen McNally*
cooking time: 45 minutes   *Orka*
cooling time: 1 hour
serves: 6-8

4 firm, ripe pears, peeled, cored   **⅔ cup firmly packed brown sugar**
   and sliced (2½ pounds)   **½ cup (1 stick) butter, cut into**
12 ounces cranberries   **pieces**
⅓ cup sugar   **½ cup all-purpose flour**
½ tsp. cinnamon   **Pinch of salt**
¾ cup rolled oats

Preheat oven to 375°. Toss pears, cranberries, sugar and ¼ tsp. cinnamon in 10-inch round baking dish until blended. Combine oats, brown sugar, butter, flour, salt and remaining ¼ tsp. cinnamon in large bowl until mixture resembles coarse meal. Sprinkle over fruit mixture and pat down lightly. Bake until pears are tender and topping is golden, about 45 minutes. Cool 1 hour. Serve warm. Vanilla ice cream or whipped cream are great partners for this old fashioned dessert.

STRAWBERRY FANTASY

preparation time: 20 minutes   *Pernilla Stahle*
serves: 4-6   *Adventure III*

2 ripe avocados   **sugar, to taste**
1 box fresh strawberries   **2-3 cups champagne**

Clean the strawberries and cut in quarters. Halve avocados and scoop out pulp with a teaspoon. Put in serving size glass dessert bowls. Sprinkle with sugar and pour champagne over top. *Serve immediately.*

CREME DE MENTHE PEARS

preparation time: 15 minutes   D.J. Parker
chilling time: 1½-2 hours   Trekker
serves: 4

1 (3 oz.) pkg. lime jello   Whipped cream
¼ c. creme de menthe   1 cup boiling water
4 canned pear halves   ¾ cup juice from pears

Dissolve jello in boiling water, add creme de menthe and juice from pears. Place pear half in individual dessert dishes and cover with jello. Chill until set. *Serve with topping of whipped cream and a drop of creme de menthe.*

STRAWBERRY ORANGE COMPOTE

preparation time: 20 minutes   Terry Boudreau
chill: 1-2 hours   High Barbaree
serves: 6

5 Navel oranges   3 Tblsp. sugar
1½ pints strawberries   2 Tblsp. orange flavored liqueur

In a ceramic or glass bowl combine 5 navel oranges, peeled, pits and membranes removed, and cut into sections. Hull the strawberries and add to orange sections with the sugar and liqueur. Transfer the compote to a container and chill covered for 1 to 2 hours.

SHARI'S STRAWBERRY CHANTILLY

preparation time: 15 minutes   Shari Stump
serves: 6   Eyola

1 Pound Cake   2 Tblsp. powdered sugar
1 pint strawberries   4 Tblsp. rum
1 pint whipping cream   ½ cup grated chocolate

Whip cream and flavor with powdered sugar and rum. Fold chocolate and strawberries left whole into mixture. *Serve over pound cake with whipped topping and grate more chocolate on top with a strawberry.*

STRAWBERRIES IN SWEDISH CREAM

preparation time: 5 minutes   *Abbie Boody*
serves: 8   *Ductmate I*

2 baskets strawberries   **3 cups sour cream**
2 cups brown sugar

Slice strawberries and layer in wine glass with sour cream, brown sugar
and strawberries. Repeat 3 times. Can prepare ahead or right before serving.

GUY'S DESSERT

preparation time: 5 minutes   *Guy B. Howe*
chilling time: 2 or more hours   *Goodbye Charlie*
serves: 6-8

8 oz. Cool Whip   **Coconut as desired**
1 pkg. pistacchio pudding mix   **½ cup small marshmallows**
½ can crushed pineapple

Mix and chill.

CHOCOLATE MOUSSE

Preparation time: 10 minutes   *Debbie Olsen*
chilling time: 1 hour   *Kingsport*
serves: 10

2 envelopes unflavored gelatin   **1 cup cocoa**
¼ cup cold water   **1 Tblsp. vanilla**
½ cup boiling water   **4 cups heavy cream, very cold**
2 cups sugar

Soften gelatin in cold water 1 minute. Add boiling water and stir until
mixture is clear. In a very cold mixing bowl combine all other ingredients
and beat until frothy. Add gelatin and spoon into parfait cups. Chill one
hour. *Serve with whipped cream and dessert cookies. This is absolutely
sinful and very easy to prepare.*

FRENCH CHOCOLATE MOUSSE

preparation time: 20 minutes   *Bobbie Gibson*
chilling time: 3 hours   *Gibson Girl*
serves: 8

4 egg yolks
¾ cup sugar
⅛ cup orange liqueur
1 (6 oz.) pkg. semi-sweet
**   chocolate pieces**
1 tsp. instant coffee, dissolved in
**   ¼ cup water**

½ cup soft butter (or margarine)
¼ cup finely chopped candied
**   orange peel**
4 egg whites
¼ tsp. salt
1 Tblsp. sugar
Whipped cream

Top of double boiler beat egg yolks and ¾ cup sugar, until lemon colored. Beat in orange liqueur. Cook over hot (not boiling) water beating constantly till thickened (10 minutes). Remove top boiler to pan of cold water and beat to consistency of mayonnaise (4-5 minutes). Melt chocolate over hot water, remove from heat, beat in dissolved coffee. Gradually add butter, beating smooth. Add peel. Stir in egg mixture. Beat egg whites to soft peaks (with ¼ tsp. salt). Add 1 Tblsp. sugar and beat stiff. Fold in chocolate-egg mixture. Pour into souffle cups, petit pots or small sherbets, cover and chill at least 3 hours. *Serve topped with whipped cream. Decorate with chrystallized violets, if you have a friend in Paris who will ship them for you.*

QUICK CHOCOLATE MOUSSE

preparation time: 15 minutes   *Terry Boudreau*
cooking time: 10 minutes   *High Barbaree*
serves: 6

9 oz. semisweet chocolate bits
½ cup hot strong coffee
3 beaten eggs

1 Tblsp. and 1 tsp. rum
1½ cups heavy cream

In the top of a double boiler blend 9 oz. semisweet chocolate bits and ½ cup hot strong coffee until the chocolate is melted, add beaten eggs and add 1 Tblsp. and 1 tsp. rum and blend the mixture until it is smooth. In a chilled bowl whip 1½ cups heavy cream until it holds stiff peaks and fold in the chocolate mixture. Transfer the mousse to dessert glasses, chill it for 30 minutes, or until it is firm, and garnish it with additional whipped cream or a sprinkling of grated chocolate.

RUM CHOCOLATE MOUSSE

preparation time: 15 minutes   *Chris Kling*
chilling time: 2 hours   *Sunrise*
serves: 6

1½ pkgs. (9 oz.) chocolate chips   1 tsp. vanilla
¼ cup dark rum   ⅓ cup sugar
3 whole eggs plus 3 yolks   1½ cups heavy cream

Place eggs, yolks, vanilla and sugar in the blender and blend for 2 minutes. Meanwhile melt the chocolate chips in the rum either over low heat or in the microwave. Pour the cream into the blender and blend for 1 minute. Add the melted chocolate and blend until smooth. Pour into individual serving cups and chill at least 2 hours.

RUM/GRAND MARNIER CHOCOLATE MOUSSE

preparation time: 30 minutes   *Ann Doubilet*
chilling time: 1-2 hours   *Spice of Life*
serves: 7-9

¼ cup sugar   3 Tblsp. whipped cream
4-6 Tblsp. rum   2 cups whipped cream
2-3 Tblsp. Grand Marnier   2 egg whites, stiffly beaten (can
¼ lb. semi-sweet chocolate   use salt or cream of tartar)

Mix rum and sugar and Grand Marnier. Cook stirring over low heat until dissolved. Melt chocolate in double boiler. Stir in 3 Tblsp. of whipped cream. Add the rum syrup and stir until smooth. Cool this mixture and add in beaten egg whites. Taste to see if you need more Grand Marnier. Fold gently into whipped cream and place in covered plastic bowl in ice box or refrigerator. *Serve in small glass bowls or wine glasses.*

BLENDER COCONUT CUSTARD PIE

preparation time: 10 minutes   *Launa Cable*
cooking time: 45 minutes   *Antiquity*
serves: 6-8

2 cups milk   4 eggs
¾ cup sugar   ¼ cup margarine
⅔ cup Bisquick   ¾ cup shredded coconut
1 tsp. vanilla   dash nutmeg

Blend all ingredients for 5 minutes in blender. Pour into greased pie pan. Bake at 350° for 45 minutes.

NEW ZEALAND PAVLOVA

preparation time: 20 minutes   *Louella Holston*
cooking time: 1 hour   *Eudroma*
serves: 6-8

4 egg whites   ¼ tsp. vanilla
1 cup sugar   1½ cups heavy cream
1 Tblsp. cornstarch   ¼ cup sugar
2 tsp. white vinegar   Your favorite fresh fruit

Beat egg white until foamy. Add the sugar a little at a time beating until meringue is very stiff. Beat in the cornstarch, vinegar, and vanilla. Spoon the meringue onto a cookie sheet lined with parchment paper and spread into a 9-inch round. Hollow out center slightly. Bake in a preheated 300° oven for 1 hour. Cool, then remove from paper onto a serving plate. Whip the cream and sugar until stiff. Fold in some of the fruit. Spread whipped cream on top of meringue. Decorate with remaining fruit.

CAY LIME PIE

preparation time: 25 minutes   *Katie MacDonald*
cooking time: 10-15 minutes   *Oklahoma Crude II*
chilling time: 3 hours
serves: 6-8

Filling: ¼ cup sugar
4 egg yolks   1 tsp. vanilla
½ cup lime juice   Gratings from 1 lime
14 oz. sweetened condensed milk   1 yummy pie crust (recipe follows)

Meringue:
4 egg whites

Beat egg yolks until soft and lemon colored. Beat in sweetened condensed milk and lime juice. Beat with electric mixer for 2-3 minutes or by hand a long time. Pour in cooled 9-inch pie crust. Beat egg whites until nearly stiff.* Add sugar and vanilla while beating a few seconds longer (not too much). Pour over filling. Make peaks with spoon and sprinkle grated lime over top. Chill as long as possible before serving. *The fresher eggs make a stiffer meringue.

YUMMY PIE CRUST

1 cup flour   ¼ tsp. salt
¼ cup brown sugar   4-5 Tblsp. cold milk
⅓ cup butter (softened)

Mix flour, brown sugar, butter and salt with fork or pastry blender. Add milk little by little. Form into ball of dough and chill if possible before rolling out. Bake at 425° for 10-15 minutes.

CHART HOUSE MUD PIE

preparation time: 20 minutes *Debbie Olsen*
chilling time: 10 hours *Kingsport*
serves: 8

½ pkg. Nabisco chocolate wafers
½ cube butter, melted
1 gallon coffee ice cream
1½ cups fudge sauce

Garnish:

Whipped cream
Slivered almonds

Crush wafers and add butter, mix well. Press into a 9-inch pie plate. Cover with soft coffee ice cream. Put into freezer until ice cream is firm. Top with cold fudge sauce (it helps to place in freezer for a time to make spreading easier). Store in freezer approximately 10 hours. *To serve: place on a chilled dessert plate and top with whipped cream and slivered almonds.*

CHOCOLATE ECLAIR PIE

preparation time: 10 minutes (plus chilling) *Samantha Lehman*
cooking time: 15 minutes *Lady Sam*
serves: 6

1 pie shell
1 box vanilla instant pudding (or
    cooked, if desired)

1 pkg. chocolate frosting mix

At breakfast, stick your pie shell in the oven and bake. Whip vanilla pudding mix until stiff, place in pie shell and refrigerate until set. Mix chocolate frosting mix (or canned) and spread over pudding mix, refrigerate until served. Is fast, easy and a successful dessert, but very rich, so easily serves 6.

PINEAPPLE CHEESE PIE

preparation time: 10 minutes *Fran Bryson*
chilling time: 1½ hours *Amoeba*
serves: 6-8

1 (8 oz.) cream cheese
¼ cup sugar
1 cup heavy cream (whipped)

1½ cups crushed pineapple
Graham cracker pie shell

Whip softened cream cheese and sugar, fold in whipped cream, then add well-drained pineapple. Pour into graham cracker pie shell. Chill thoroughly and serve.

LEMON CRUNCH PIE

preparation time: 45 minutes
chilling time: 2-3 hours
serves: 6

Margaret Benjamin
Illusion II

½ lb. ginger cookies
3 oz. butter or margarine
8 oz. condensed milk

2 lemons
¼ pt. whipping cream
Grated chocolate

Crumble ginger cookies between wax paper with rolling pin. Melt butter and blend with ginger crumbs. Press down in the bottom of a 9-inch pie dish. Put into refrigerator until set. Pour condensed milk into bowl, add the rind and juices of the lemons and stir until it thickens. Whip cream until thick and standing in peaks and then fold into lemon mixture. Pile into ginger base and grate chocolate on top for decoration. (This can be made with a graham cracker crust but it doesn't taste as good.)

MANGO PIE

preparation time: 30 minutes
cooking time: 1 hour 10 minutes

Betsy Sackett
Insouciant

1½ cups flour
1 tsp. salt
¼ tsp. baking powder

1 tsp. sugar
½ cup wesson oil
3 Tblsp. milk

Sift dry ingredients into pie pan. Add milk to oil, beat well. Mix with dry ingredients. Pat out into the pie pan.

3 cups sliced mangoes
¼ cup sugar
Juice of ½ lemon

1 tsp. cinnamon
½ tsp. nutmeg
½ tsp. allspice

Mix and put into pie shell.

Topping:

¾ cup flour

½ cup sugar
⅓ cup butter

Mix until crumbly, sprinkle on top of mangoes. Bake at 425° for 10 minutes. Turn to 350° for about an hour until done.

• *To prevent soggy pie crust from juice of the filling, brush crust with egg white.*

KEY RUM PIE

preparation time: 30 minutes
cooking time: 5 minutes
chilling time: 2½ hours
serves: 6

Casey Wood
Megan Jaye

½ cup Mt. Gay, white rum
1 pkg. lemon pudding
1 pkg. lime jello
⅓ cup sugar
2½ cups water

2 eggs, slightly beaten
1½ cups whipped cream or Cool
   Whip
1 graham cracker crust
Lime slices

Mix pudding, jello and sugar in a saucepan. Blend in ½ cup water and eggs. Add remaining water and cook over medium heat until boiling. Remove. Stir in rum. Chill. Fold in whipped cream, spoon into pie crust and chill until firm, about 2 hours. Garnish with lime slices.

PEANUT BUTTER PIE

preparation time: 10 minutes
chilling time: 2 hours
serves: 8

Debbie Olsen
Kingsport

Pie:
1 (8 oz.) pkg. cream cheese,
   softened
1 (14 oz.) can Eagle sweetened
   condensed milk (not
   evaporated milk)
1 cup peanut butter

1 (4 oz.) Cool Whip, thawed
1 tsp. vanilla
1 (6 oz.) jar Smuckers Chocolate
   Fudge Sauce
1 Chocolate Crunch Crust, recipe
   follows

In large bowl, beat cream cheese until fluffy, beat in condensed milk, peanut butter and vanilla until smooth. Fold in whipped cream. Turn into crust. Top with fudge sauce (it helps to place in freezer for a time to make spreading easier). Chill in freezer for about two hours. This is delicious and similar to a peanut butter cheesecake.

CHOCOLATE CRUNCH CRUST:

In heavy saucepan, over low heat, melt ⅓ cup butter and 1 small pkg. (6 oz.) chocolate chips. Remove from heat and gently stir in 2½ cups Rice Krispies cereal until completely coated. Press into bottom and sides of greased 9-inch pie plate. Chill 30 minutes.

PINEAPPLE SOUR CREAM PIE

preparation time: 30 minutes   *Liz Thomas*
chilling time: 3 hours   *Raby Vaucluse*
serves: 6

1 pkg. instant vanilla pudding and   2 cups sour cream
   pie filling   1 Tblsp. sugar
1 can (8 oz.) crushed pineapple   1 (9-inch) baked pie shell

Combine pie filling mix; pineapple and juice, sour cream and sugar. Beat with rotary beater for 1 minute. Pour in pie shell. Chill about 3 hours.

RASPBERRY CLOUD PIE

preparation time: 10 minutes   *Ann Glenn*
chilling time: 2 hours   *Encore*
serves: 6

1 pkg. raspberry jello (3 oz.)   8 oz. Cool Whip
⅔ cup boiling water   1 Graham Cracker Ready Crust
10 oz. pkg. quick-thaw
   raspberries

Dissolve gelatin in boiling water. Stir well. Add frozen raspberries and stir well. Blend in Cool Whip and whisk till smooth. Spoon into pie crust and chill 2 hours.

SOUR CREAM RAISIN PIE

preparation time: 15 minutes   *Lori Moreau*
cooking time: 45-50 minutes   *Alberta Rose*
serves: 6

1 cup sugar   ½ tsp. nutmeg
1 egg   ½ tsp. cinnamon
1 cup chopped raisins   Dash ground cloves
1 cup sour cream   1 9″ pie shell, unbaked

A very simple pie and very delicious. Combine all filling ingredients with a wire whip and blend well. Pour into pie shell and bake at 400° for 15 minutes, then 30 minutes at 350°. *Serve hot or cold.*

SOUTHERN PECAN PIE

preparation time: 15 minutes   *Jeannie Drinkwine*
cooking time: 45 minutes   *Vanda*
serves: 8

1 unbaked 8-inch pastry shell
3 eggs, whole, slightly beaten
1 cup Karo, dark corn syrup (blue
    label)
⅛ tsp. salt
1 tsp. vanilla

2-3 drops almond extract
1 cup sugar
2 Tblsp. melted butter
2 cups pecans, chopped (save
    some whole ones to decorate
    top of pie)

Mix eggs, Karo syrup, salt, vanilla, almond, sugar and butter. Stir in pecans. Pour in shell. Bake in hot oven (400°) for 15 minutes; reduce heat to 350° and bake 30 to 35 minutes longer. Filling should appear slightly less set in center.

TRIFLE

preparation time: 15 minutes   *Alison P. Briscoe*
cooking time: 15 minutes   *Ovation*
chilling time: 3 hours
serves: 6

1 packaged pound cake (small)
1 cup cream sherry
1 small jar strawberry preserves
6 eggs
2 Tblsp. cornstarch
½ pt. (3 small cartons) whipping
    cream
2 Tblsp. fine sugar

2 Tblsp. milk
4-5 drops vanilla essence
1 can Kraft Dairy Cream (Redi
    Whip)
Chopped walnuts
Chopped almonds
Chopped cherries

Place sliced pound cake on bottom of deep dessert dish. Soak with sherry and spread with preserves. Chill for 2 hours. Sieve cornstarch. Mix to a paste with milk in a bowl. In saucepan bring cream to almost boiling point. Remove from heat. Beat eggs and mix into cornstarch paste. Add sugar. Add hot cream to this, slowly, whisking thoroughly. Reheat this mixture in saucepan whisking until thick. Add vanilla. Pour onto chilled pound cake and chill for 1 hour. Spread Redi Whip cream on top. Sprinkle with nuts and cherries.

HOT BROWNIE SOUFFLE WITH SAUCE

preparation time: 35 minutes   *Kathleen McNally*
cooking time: 25 minutes   *Orka*

Butter
Sugar
½ cup butter (1 stick) chopped
4 ounces unsweetened chocolate,
    coarsely chopped
1 cup sugar
4 egg yolks, room temperature

1 Tblsp. instant coffee powder,
    dissolved in 1 Tblsp. rum or
    orange liqueuer. I prefer
    orange liqueur.
1 tsp vanilla
¼ cup all-purpose flour
5 egg whites, room temperature
Vanilla Ice Cream Sauce, recipe
    follows

Position rack in middle of oven. Preheat to 450°. Butter souffle dish and sprinkle with sugar. Melt ½ cup butter with chocolate over low heat, stirring until smooth. Blend in ¼ cup sugar, yolks, coffee mixture and vanilla. Stir in flour. Beat whites in large bowl until soft peaks form. Gradually add remaining ½ cup sugar beating constantly until whites are stiff, but not dry. Fold ¼ of whites into chocolate, then fold chocolate back into remaining whites (be careful not to deflate mixture; a few streaks of white may remain). Turn batter gently into prepared dish. Sprinkle lightly with sugar. Bake 5 minutes. Reduce oven temperature to 400° and continue baking until souffle is puffed, about 20 minutes (center will remain moist). Serve immediately with sauce.

VANILLA ICE CREAM SAUCE:
Makes 2 cups

1 pint rich vanilla ice cream   2 Tblsp. rum or orange liqueur

Place ice cream in medium bowl. Let soften at room temperature. Add liqueur and beat until smooth. Turn into small bowl and serve immediately with souffle. *Tastes like a rich super moist brownie — Fantastic!!!*

SORBET A LA BRUXELLES

preparation time: 15 minutes   *Betsi Dwyer*
serves: 6-8   *Malia*

1 pint raspberries or strawberries   1 quart vanilla ice cream
Brandy or Kirsch, to taste (about
    ¼ cup)

Crush berries and press through a strainer to remove the seeds. Sweeten to taste and flavor with brandy. Stir into slightly softened ice cream and serve in dessert glasses. This dessert should not be firm.

COFFEE SOUFFLE

preparation time: 40 minutes   *Margaret Benjamin*
chilling time: 1-2 hours   *Illusion II*
serves: 6

4 eggs
4 oz. sugar
1½ Tblsp. instant coffee, dissolved
   in 6 Tblsp. boiling water

½ pt. whipping cream
1 packet gelatin, dissolved in 5
   Tblsp. water

Separate the eggs, put the yolks, sugar and coffee in a bowl and beat until thick. Dissolve the gelatin in the 5 Tblsp. water over a low heat. Beat
the egg whites in bowl until stiff and peaking. Beat the cream until
whipped. Add the gelatin to the yolk mixture and then fold in the white
of egg and then the cream. Pour into a souffle dish and put in refrigerator
until set.

VANILLA PARFAIT

preparation time: 15 minutes   *Pernilla Stahle*
cooking time: 10-15 minutes   *Adventure III*
freezing time: 3-4 hours
serves: 6

4 egg yolks
½ cup confectioners sugar
1 vanilla stick (or substitute)
¾ cup heavy cream

6 meringue bottoms
6 meringue tops
3 kiwi fruits
1 can blackberries

Scrape out the middle of the vanilla stick, or take vanilla extract and put
in a saucepan. Add egg yolks and sugar. Put saucepan in boiling water,
whip until thick. Whip cream until thick and add. Divide mixture into portion forms, and freeze 3-4 hours or overnight. Put meringue bottom on
plate, then the parfait, meringue on top. Slice kiwi on the side and 2
Tblsp. blackberries on each plate. Note: Use vanilla bean ice cream if not
time enough.

EASY CHERRY TARTS

preparation time: 5 minutes   *Linda Richards*
cooking time: 15-20 minutes   *Tao Mu*

2 cans cherry pie filling
8-10 tart shells

1 pt. whipping cream or canned
   cream

Preheat oven to 375°. Fill tart shells with cherry filling. Place in preheated oven on cookie sheet or muffin tins until cherry fillings are just heated
through. Add cream over top just before serving.

WALNUT TART

preparation time: 20 minutes   Nancy Thorne
cooking time: 30 minutes   Gullviva
serves: 6

Shell: **Filling:**
⅓ cup butter   2 cups walnuts, coarsely chopped
¼ cup sugar   ⅔ cup light brown sugar, packed
1 egg yolk   ¼ cup butter
1 cup flour   ¼ cup dark corn syrup
                             2 Tblsp. heavy cream
                             Garnish: ½ cup heavy cream

Shell: Beat butter, sugar and egg yolk. Beat in flour just until blended. Preheat oven to 375°. Press dough into bottom and sides of 9 inch tart pan. Bake 12 minutes or until lightly browned. Cool. Bake nuts on cookie sheet 5 minutes.

Stir sugar with butter, corn syrup and 2 Tblsp. heavy cream. Boil one minute. Put walnuts in shell. Pour mixture over walnuts. Bake 10 minutes or until bubbly. Beat whipped cream and serve on side.

WALNUT TORTE

preparation time: 30 minutes   Barbara Tyne
cooking time: 30 minutes   Anodyne
chilling time: 2 hours
serves: 8

3¼ cup walnuts   6 eggs, separated, at room
3 Tblsp. flour   temperature
1 tsp. baking powder   1 cup sugar
¼ tsp. salt   1 cup heavy cream
                             2 tsp. sugar

Grease 2 8-inch cake pans, line with wax paper and grease the paper. Coarsely chop 1 Tblsp. walnuts and set aside for garnish. Finely chop the remaining walnuts and put in a bowl along with the flour, baking powder and salt. Beat egg whites until soft peaks form. Beat egg yolks with 1 cup sugar until light and lemon colored. Gently fold nut mixture into whites and then fold in yolk mixture. Spoon into cake pans and bake for 25-30 minutes in a preheated 350° oven. Allow to cool in pan for 5 minutes, turn out onto cooling racks and remove paper. Cool completely. Whip cream with remaining sugar until firm. Spread cream on top of each layer and stack on a serving plate. *Refrigerate at least 2 hours before serving and garnish with reserved walnuts.*

RECIPE FOR A HAPPY DAY

preparation time: none
cooking time: none
serves: lots

Michelle Sutic
Harp

1 cup friendly words
2 heaping cups understanding
4 heaping tsp. time and patience

Pinch of warm personality
Dash of humour

Measure words carefully. Add heaping cups of understanding. Use generous amounts of time and patience. Cook with gas on front burner. Keep temperature low — do not boil. Add dash of humour and pinch of warm personality. Season to taste with spice of life. *Serve in individual molds.*

PARTICIPATING YACHTS

Adventure III: *Captain Mats Lovkvist, Chef Pernilla Stahle*
Have more fun on a Swedish Swan 51'.
We don't just sail well, we fly on top of the water.
We don't just eat OK, we excel in gourmet delights.
We don't just have an enjoyable time, we make sure it is
your best vacation ever!

Alberta Rose: *Captain Gene Moreau, Chef Lori Moreau*
Alberta Rose, a semi-custom Morgan 51' chartering since 1974 and since 1979 in the
Virgins, accommodates 4 guests and 2 crew with 3 private, double staterooms.
Although luxurious and comfortable below deck, Alberta Rose does not lack in sailing
ability and is a beautiful platform for seeing the Virgin Islands.

Amoeba: *Captain R.E. Bryson, Chef Fran Bryson*
54 foot schooner; maximum 6 guests.

Anemone: *Captain Blair Albert, Chef Cindy Glover*
Anemone is a Nautical 56' Ketch based in St. Thomas. She is in bristol condition
and boasts 3 separate staterooms, each with own head and vanity, separate crews
quarters and 3 full showers below. We carry a windsurfer, snorkle gear, water skis,
fishing gear and 13 foot Boston Whaler with 35 HP engine.

Ann-Marie II: *Captain Ken Haworth, Chef Barbara L. Haworth, R.D.*
This 37 foot steel Zeeland Yawl was built in Holland over 25 years ago and has an
interior of mahogany and stain-leaded glass. Ken and Barbara Haworth invite you to
join them for a sail in the islands. Enjoy Barbara's gourmet creations and the couple's
warm hospitality.

Anodyne: *Captain Tom Tyne, Chef Barbara Tyne*
Anodyne, a 37 foot center cockpit Sloop, is owned and operated by Tom and
Barbara Tyne. The spacious aft cabin allows Anodyne's two guests, privacy
unprecedented on yachts this size. Barbara's background as magazine food editor,
restaurant reviewer, cooking school teacher and food consultant make the meals on
board a delicious adventure.

Antipodes: *Captains Dieter and Manfred Zerbe, Chef Silvia Kahn*
It's a pleasure to introduce my international meals and share menus which I have
mastered in my galley. I was born in Venezuela of Austrian and Turkish descent
Extensive travel has enabled me to experience a variety of dishes. "Buen Provecho"
Antipodes: 60 foot Schooner.

Antiquity: *Captain David Decuir, Chef Launa Cable*
41 foot Ketch; maximum 5 guests.

Avatar: *Captain Harvey Beberman, Chef Harriet Beberman*
Avatar, luxurious 46 foot Motor Yacht, by the master builders of Cheoy Lee, is fully
equipped with every modern convenience including air conditioning, roll stabilizer and
radar. Waterski, windsurf, fish, or dive. Memorable dining experience.
Owner/operated by Harvey and Harriet Beberman.

Bagheera: *Captain and Chef Joel H. Jacobs*
True gourmet cooking in an elegant setting of fine furnishings, objects de art and antiques on board a vessel that can do 200 miles per day under sail alone. Your assurance is our many letters complimenting us on the fine food, furnishings, and attentiveness of our crew. The elegant vacation!

Barefoot: *Captain John Blythe, Chef Phoebe Cole*
For a wonderful vacation come "Barefooting" with us, John and Phoebe.

Bon Papa D: *Captain Sune Ostlund, Chef Ci Ci*

Cantamar IV: *Captain Don Stone, Chef Maureen Stone*
Cantamar IV, an Ocean 60 owned and operated by Don and Maureen Stone, is a fast, roomy schooner accommodating up to eight guests. Don and Maureen sailed their 40 foot Ketch from Canada seven years ago and have been chartering in the Virgins ever since.

Caribe Monique: *Captains Leayle Benjamin and Erica Benjamin*
57 foot Chris Craft Power Boat

Carriba: *Captain Clayton Ballantine, Chef Kathleen Leddy*
46 foot Morgan.

Coyaba: *Captain Brian Fleming, Chef Ginger Outlaw Fleming*

Daddy Warbucks: *Captain Brent Hamilton, Chef Trish Penn*
Sail in comfort on this luxurious Irwin 52. Incomparable cuisine by newly published, Trish. Virgin Islands or Down Island flexibility is our keynote. Accommodates up to six in three separate stateroooms each with head, two showers Windsurf, snorkel and VHS. Most reasonable price in size range. Come sail with us!

Different Drummer: *Captain Phil Clarkson*
50 foot Ketch.

Ductmate: *Captain Brian Fulford, Chef Abbie Boody*
100 foot Power Boat; maximum 8 guests.

Encore: *Captain Marvin Glenn, Chef Ann Glenn*
Encore's 50' long triple hulls provide spaciousness for up to eight guests on deck and below. Her 27' width provides stability under sail and at anchor. Owner/operators, Ann and Marvin Glenn, did an 8 yeart world cruise on a smaller Trimaran before buying Encore in 1977 for chartering.

Eudroma: *Captain Tim Holsten, Chef Louella Holsten*
Eudroma, 78 feet of classic elegance and comfort sails you to adventurous times, exciting anchorages to treasure hunt, windsurf, snorkel, sightsee, or shop. Experience gourmet cuisine prepared and served by Eudroma's owners and operators.

Eyola: *Captain Hennie Page, Chef Shari Stump*
Eyola is a 90' Brigantine built by owner/operator Hennie Page. Her spacious interior consists of three double cabins with comfortable double beds. Enjoy exquisitely presented and tastefully prepared meals which are a specialty of Eyola. Dine by candle light in our beautifully decorated salon. Your deserve this luxurious vacation!

Farley: *Captain Bill McKay, Chef Sue McKay*
   47 foot Sloop; maximum 4 guests.

Gibson Girl: *Captain Bob Gibson, Chef Bobbie Gibson*
   49 foot Motor Yacht.

Glen Mac: *Captain John Booey, Chef Terry Booey*
   Congenial owner/operated - two double cabins, each with private head and shower. Spend your vacation the way you want to; sailing, diving (compressor onboard), windsurfing, sightseeing, sunning or just plain relaxing with TV, video, and large library. Your vacation ideas become reality on a comfortable yacht, adaptable to your wants.

Goodbye Charlie: *Captain Guy Howe*
   55 foot luxury Long Range Trawler. Spacious and comfortable oversized lounge with separate galley up. Lounge bar with ice maker and refrigerator. Three private staterooms and two heads. Separate crews quarters. Four separate reverse cycle air conditioners. Large swim platform. Captain graduate of Maine Maritime.

Great Escape: *Captains Jay Wilbur and Kathi Strassel, Chef Kathi Strassel*
   Great Escape, a 54 foot Ketch designed by Bob Perry is a fast and lovely world cruiser outfitted for total comfort. There are three private, double staterooms for up to six guests, separate crew quarters and a galley totally equipped to provide the best in dining pleasure.

Grumpy III: *Captain Charlie Peet, Chef Marty Peet*

Gullviva: *Captain John Wolff, Chef Nancy Thorne*
   Gullviva, a well kept and artistically decorated Gulfstar 50', being run by a young and active team, John and Nancy. Both have had extensive charter boat experience in the Virgin Islands; John is a dive master and marine life enthusiast and Nancy has spent ten years as a professional gourmet chef.

Harp: *Captain Krasnic Sutic, Chef Michelle Sutic*
   This Gulfstar 60 commissioned especially for chartering has expansive teak decks for roaming and relaxing, cushioned aft deck for lounging and large center cockpit shaded by a bimini for dining alfresco. Below deck, the main salon seats 6 around a beautiful teak table. Three private cabins accommodate guests. Air conditioning throughout.

High Barbaree: *Captain Walter Boudreau, Chef Terry Boudreau*
   High Barbaree is a 78 foot Ketch designed by Phillip Rhodes. Formerly owned by Don Juan of Spain, she is now a luxurious charter yacht with accommodations for up to 6 guests and carries a crew of 5. The food is "scrumptious". High Barbaree cruises the Virgins and Windwards.

Illusion II: *Captain Roger Perkins, Chef Margaret Benjamin*
   Try the exhilaration and freedom of sailing Illusion II, a 53 ' custom sailing Ketch. Have fun windsurfing, snorkeling, and sunbathing on her spacious flush deck. Below deck she accommodates up to six people in the greatest comfort. Roger and Margaret, your English captain and chef promise you an unforgettable vacation.

Insouciant: *Captain Steve Young, Chef Betsy Sackett*
   48 foot Yawl; maximum 5 guests.

Kingsport: *Captain Roy Olsen, Chef Debbie Olsen*
Kingsport, an Irwin 65 Ketch, has four double cabins including exceptional master stateroom and beautiful, spacious main salon. Specializing in scuba diving, water skiing, windsurfing, and the Captain's favorite "Banana Coladas". Kingsport has two VHS systems with over 100 taped movies and other indulgences.

Lady Sam: *Captain George Lehman, Chef Samantha Lehman*
Gosh, come to my galley. I do not measure but gain great knowledge in the taste test. "A dash of this and that makes for the superb." Our guests enjoy my company so I cut all corners in the galley to be sociable. So far, total success and satisfaction.

Malia: *Captain Eric Unterborn, Chef Betsi Dwyer*
This 53' fiberglass Ketch was designed by Pearson Yachts for chartering. Eric and Betsi welcome you aboard in St. Thomas during the winter and Newport, R.I. during the summer. Sail, swim, windsurf, fish or relax in the spacious cockpit. Dine on classic cuisine from five continents. It's the greatest vacation.

Maranatha: *Captain Rob Fortune, Chef Shirley Fortune*
54 foot Ketch.

Megan Jaye: *Captain Tom Miller, Chef Casey Wood*
43 foot Sloop; maximum 4 guests.

Mystique: *Captain Ken Culbert, Chef Cassandra (Sandy) Culbert*
Ken and Sandy take from 2 to 6 persons on their 50 foot yacht Mystique for a week or more cruise of the Virgin Islands. Sailing, snorkeling, exploring, scuba diving and just relaxing. Gourmet meals.

Nani Ola: *Captain George Lang, Chef Jessica Adam*
A large 48 foot Ketch with a unigue layout designed to sleep 4 guests in comfort and privacy. Two heads with showers, main dining salon and a unique lower level lounging area for both before and after dinner cocktails or just visiting. Fully equipped for your safety and pleasure.

Nordic: *Captain Deek Vernon, Chef Juli Vernon*
Deek and Juli Vernon probably have more sailing and charter experience between them than most professional sailors taking fair advantage of the wind. Nordic, their big bold, and beautiful 83 foot Yawl, is for sailing aficionados demanding the utmost in safety, comfort and luxury under sail or at anchor in a quiet cove.

Oklahoma Crude II: *Captain Trevor Rees, Chef Katie MacDonald*
Oklahoma Crude II is an Erwin 52 with all the luxuries of home including A.C. w/w carpet, colored TV, icemaker and windsurfers. On top of all this, she sails like a dream. As for the crew, we love what we do. Sail with us. You will love it too!

Orka: *Captain Kris Van de Vyrer, Chef Kathleen McNally*
Join Kris and Kathy aboard Orka, a modern 56' Ketch. Sailing, scuba diving, water skiing, windsurfing, spinnaker flying, snorkeling, and fishing. Full scuba diving equipment, compressor and dive master. 12 years charter experience, 4 years Virgin Islands. An active vacation for active people. Gourmet cuisine. Orka takes up to 7 guests.

Ovation: *Captain Douglas Briscoe, Chef Alison P. Briscoe*
Ovation is a splendid sailing yacht! An Ocean 60 designed to be the most comfortable, true ocean sailing charter yacht her size in the industry today. Featuring four equal, double cabins for up to eight guests. We would love to have you as our guests aboard Ovation - Bon Appetit!

Paradise: *Captain Doug Thayer, Chef Jean Thayer*
In a 50 foot Gulfstar "Paradise" combine 2 or more people. Over medium-high heat (82 degrees) sail for 1 week or more, mixing frequently with beaches, snorkeling and swimming. Blend with Jean and Doug Thayer; meltingly good meals; dashes of favorite libations; add sun, sand and sea to taste. It's Paradise!

Pieces of Eight: *Captain Ray McMichael, Chef Pamela McMichael*
Pieces of Eight is a very classy and comfortable Irwin 65'. Ray and Pam McMichael are a fun loving couple who will be your hosts around the islands.

Plantos: *Captain Thomas Nunn, Chef Kathy Wagner*

Polaris: *Captain David Cook, Chef Gill Case*
Polaris is a 55' luxury Motor Yacht with three guest staterooms, air conditioning, stereo, television, VHS movies and decks everywhere for sunning and reading. She offers water skiing, sailing, windsurfing, fishing, snorkeling, and superb food prepared to your exact taste. WARNING: It can be painful to return home!

Raby Vaucluse: *Captain Bill Gibson, Chef Liz Thomas*
This Morgan 46 is built for charter. She is very comfortable below with 2 guest staterooms and private heads. Bill Gibson is highly competent and experienced. Liz Thomas takes great pride in her meals prepared aboard. Let Raby Vaucluse and her crew make this your most enjoyable vacation.

Saracen: *Captain Jack du Gan, Chef Kyle Perkins*
80 foot Yawl.

Satori: *Captain Edward Thompson, Chef Gabrielle Thompson*
Ed and Gay Thompson built their beautiful 57' gaff-rigged Schooner Satori in California 1975 and have been chartering in the Virgin Islands since 1977. This spacious yacht has 7' long double or king size beds throughout. Her large galley allows Gay the freedom to prepare new masterpieces for her guests.

Silver Queen of Aspen: *Chef Cheryl Fleming*
52 foot Irwin.

S'Oublier: *Captain John Muller, Jr., Chef Carol Muller*
S'Oublier is a luxurious Cheoy Lee Offshore 44' Ketch. Lavishly appointed for one couple or family of four. Separate access to aft cabin with king size berth and shower for maximum privacy. Expansive teak decks and covered cockpit. Enjoy sailing, snorkeling, underwater camera and dinghy. Mature, capable and caring crew.

Spice Of Life: *Captain David Darwent, Chef Ann Doubilet*
58 foot Cutter; maximum 6 guests.

Sunrise: *Captain Jim Kling, Chef Chris Kling*
Sunrise, a 55' Cutter offers 4 guests spacious, private staterooms, great sailing and a classic old world ambiance below. Beautiful wood carvings, delicately inlaid tables, leaded glass cabinets; these are a few of the loving details that captain and mate, Jim and Chris Kling added as they built their special yacht.

Tao Mu: *Captain Mel Richards, Chef Linda Richards*
Tao Mu is a 48' Tri with double cabins for eight guests. Chartering successfully since 1973, she features fast and stable sailing with abundant deck space, plus excellen cuisine. Mel and Linda Richards are Captain and Hostess whose goal is your enjoyment.

Taza Grande: *Captain Barry Biggs, Chef Adriane Biggs*
Taza Grande is an Irwin 65 Ketch with accommodations for six guests in three separate cabins.

Trekker: *Captain Rob Parker, Chef D.J. (Dottie) Parker*
Fulfill your island fantasy aboard Trekker. This Bristol 40 with captain and mate cater to two guests at a time. Food, liquor, wine, snorkel gear, scuba arrangements, stereo, color TV and much more are included in the low rate of $1850 per week/couple Join us for an unforgettable holiday.

Tri-World: *Captain Lloyd Royle, Chef Jennifer Royle*
Dive, windsurf, snorkle, water ski and sail our sunfish. Dive master/instructor aboard, also scuba equipment and compressor. Three private cabins each with head, shower, wash basin and double berth. Seven years chartering and diving in the U.S. and B.V.I. An active boat for active people.

Vanda: *Captain Paul Drinkwine, Chef Jeannie Drinkwine*
Vanda, a magnificent owner/operated 93 foot Ketch. She is a classic yacht built in England in 1909 for aristocrats and available for those who expect and demand quality Luxury is her hallmark with a crew of four dressed in whites and continental cuisine served on silver, pewter and crystal.

Vanity: *Captains Bob and Jan Robinson, Chef Jan Robinson*
Vanity, a luxurious 60' Motor Sailer. This owner/operated charter yacht is designed for privacy. Taking only four guests, Vanity assures you of individual consideration Dine comfortably under the stars, enjoy international cuisine. Nine years of charterin; give the Robinsons the knowledge of where the action is and where serenity prevails.

Viking Maiden: *Captains Paul and Paula Taylor, Chef Paula Taylor*
Viking Maiden is a luxury yacht built specifically for charter, with four equal double staterooms, each with private head and shower. Paul and Paula combine British and American hospitality to welcome you on board to enjoy one of the most memorable holidays of a lifetime.

Western Star: *Captain Mel Luff, Chef Patty Dailey*
Sailing, snorkeling, scuba diving, waterskiing, windsurfing, even Jet Skis are available on Western Star's maxi fun Virgin Island charters. Sail with us aboard our 53 Gulfstar and enjoy the ultimate in cruising luxury. Think about the crystal clear waters warm sun, deluxe meals and energetic friendly crew.

Whisker: *Captain Dave Ferneding, Chef Roz Ferneding*
65 foot Power Boat; maximum 6 guests.

Wind Song Oregon: *Captain Jeffery Burger, Chef Kandy Popkes*
58 foot custom Ketch.

INTERNATIONAL IDIOMS AS TO
FOODS AND COOKING UTENSILS

AMERICAN	BRITISH
all-purpose flour	plain flour
baking pan	baking tin
baking soda	bicarbonate of soda
beets	beetroot
braising beef	stewing steak
bread flour	strong flour
broil	grill (top heat)
candied	glace'
canning	bottling
cookie sheet	baking sheet
cookies	sweet biscuits
clean	gut
confectioners' sugar	icing sugar
egg plant	aubergine
extract	essence
golden raisins	sultanas
granulated sugar	caster sugar (is a little finer)
green onion or scallion	spring onion
grill	base grilling (on barbecue)
half and half	single cream
haricots verts	French beans
heavy cream	double cream
jellies (without fruit)	jellies (without fruit)
jello, gelatin dessert	jelly
lima beans	broad beans
luke warm	blood heat
Navy beans	haricot beans
rolled oats	oats, porridge
rutabaga	swede
shredded coconut	dessicated coconut
sift	sieve
skillet	frying pan
snap beans	runner beans
zucchini	courgette

RE-ORDER ADDITIONAL COPIES

Quantity	Description	Price	Total
	Ship To Shore I	$16.95	
	Sweet To Shore	$15.95	
	Sea To Shore	$15.95	
	Sip To Shore	$12.95	
	Slim To Shore	$15.95	
	BahamaMama's Cooking	$12.95	
	6% Tax (N.C. Only)		
	Gift Wrap $2.50 Each		
	Freight $3.00/Book		
	TOTAL		

AUTOGRAPH TO: _____

SHIP TO: _____

AUTOGRAPH TO: _____

SHIP TO: _____

Call to order **TOLL FREE**
1-800-338-6072

Please charge my: ☐ VISA
☐ MasterCard

Card Number:

⬚⬚⬚⬚⬚⬚⬚⬚⬚⬚⬚⬚⬚⬚⬚⬚

Signature:

_____ _____

Expiration Date: _____ _____

Make **check** and **Money Orders** payable to:

SHIP TO SHORE, INC.
10500 MT. HOLLY ROAD
CHARLOTTE, NC 28214-9219

- -

RE-ORDER ADDITIONAL COPIES

Quantity	Description	Price	Total
	Ship To Shore I	$16.95	
	Sweet To Shore	$15.95	
	Sea To Shore	$15.95	
	Sip To Shore	$12.95	
	Slim To Shore	$15.95	
	BahamaMama's Cooking	$12.95	
	6% Tax (N.C. Only)		
	Gift Wrap $2.50 Each		
	Freight $3.00/Book		
	TOTAL		

AUTOGRAPH TO: _____

SHIP TO: _____

AUTOGRAPH TO: _____

SHIP TO: _____

Call to order **TOLL FREE**
1-800-338-6072

Please charge my: ☐ VISA
☐ MasterCard

Card Number:

⬚⬚⬚⬚⬚⬚⬚⬚⬚⬚⬚⬚⬚⬚⬚⬚

Signature:

_____

Expiration Date: _____

Make **Checks** and **Money Orders** payable to:

SHIP TO SHORE, INC.
10500 MT. HOLLY ROAD
CHARLOTTE, NC 28214-9219

PERFECT GIFTS
FOR ANY OCCASION
FREE!

Share the taste of the Caribbean with your Friends!
We will send your friends a FREE catalog.
Simply mail this form or call 1-800-338-6072.

Name: _____

Address: _____

City: _____ State: _____ Zip: _____

Name: _____

Address: _____

City: _____ State: _____ Zip: _____

SHIP TO SHORE, INC.
10500 MT. HOLLY ROAD
CHARLOTTE, NC 28214-9347

- -

PERFECT GIFTS
FOR ANY OCCASION
FREE!

Share the taste of the Caribbean with your Friends!
We will send your friends a FREE catalog.
Simply mail this form or call 1-800-338-6072.

Name: _____

Address: _____

City: _____ State: _____ Zip: _____

Name: _____

Address: _____

City: _____ State: _____ Zip: _____

SHIP TO SHORE, INC.
10500 MT. HOLLY ROAD
CHARLOTTE, NC 28214-9347